Statistics for Analytical Chemists

——STATISTICS FOR——
ANALYTICAL CHEMISTS

Roland Caulcutt

and

Richard Boddy

Statistics for Industry (UK) Ltd

LONDON   NEW YORK

CHAPMAN AND HALL

First published 1983 by
Chapman and Hall Ltd
11 New Fetter Lane, London EC4P 4EE
Published in the USA by
Chapman and Hall
733 Third Avenue, New York NY10017

© *1983 R. Caulcutt and R. Boddy*

Printed in Great Britain by
J. W. Arrowsmith Ltd, Bristol

ISBN 0 412 23730 X

British Library Cataloguing in Publication Data

Caulcutt, Roland
  Statistics for analytical chemists.
  1. Mathematical statistics
  I. Title   II. Boddy, Richard
  519.5′0246541   QA276

  ISBN 0-412-23730-X

Library of Congress Cataloging in Publication Data

Caulcutt, R. (Roland)
  Statistics for analytical chemists.

  Bibliography: p.
  Includes index.
  1. Chemistry, Analytic—Statistical methods.
I. Boddy, Richard, 1939–   . II. Title.
QD75.4.S8C38 1983   519.5′024541   82-23637
ISBN 0-412-23730-X

Contents

————Preface————

This book is based upon material originally prepared for courses run by Statistics for Industry (UK) Ltd. Over a number of years written material was repeatedly extended and refined as the authors developed a clearer image of the statistical needs of the course members. This development was retarded, however, by the adoption at the outset of an assumption that proved to be false. We assumed that the majority of those scientists sharing the title 'analytical chemist', would also share common problems and have very similar needs. We were mistaken.

Experience revealed, for example, that many analytical chemists were primarily concerned with errors of calibration, whilst others had no interest whatsoever in this area, but were seeking advice on the problems of inter-laboratory trials. We also found that this latter interest was shared by scientists and technologists of other disciplines, ranging from civil engineers to nuclear physicists.

It would be foolish, therefore, to suggest that this book offers ready solutions to *all* the statistical problems of *every* analytical chemist. We could reasonably claim, however, that *any* analyst with a statistical problem would benefit from adopting this text as a foundation for his introductory reading. The approach adopted throughout the book has two distinctive features, which also character-ize the courses on which the book is based. This approach is both 'problem centred' and 'non-mathematical', thus enabling the reader to concentrate upon three essential elements:

(a) how to use statistical techniques,
(b) how to interpret the results of statistical analysis,
(c) how to check the assumptions underlying the statistical techniques.

Many of the statisticians and scientists who lecture on Statistics for Industry courses have made numerous suggestions during the preparation of this book. These suggestions have undoubtedly led to an improvement in the readability and the usefulness of the book. We are deeply grateful to all concerned, who are too numerous to list individually. We are particularly grateful to Tony Wilson (now retired but formerly with the Water Research Association) for helping us to appreciate the diverse problems of the analytical chemist and for offering guidance in our search for solutions.

The policy of continuous improvement adopted by Statistics for Industry has hopefully, resulted in a more readable book. On the other hand this same policy does makes great demands upon our typist, Christine Robinson, who has managed to produce a beautiful typescript from the mis-spelt jottings of one author and the illegible notes of the other.

Statistics for Industry (UK) Ltd runs many courses in applied statistics at a variety of venues throughout each year. All of these courses, or *workshops* as they are more correctly described, are intended for scientists and technologists in the chemical and allied industries. The current list of workshops includes:

Introduction and Significance Testing
Statistics in Research and Development
Statistics for Analytical Chemists
Design of Experiments
Statistical Quality Control

In response to the needs of our customers new workshops are introduced every year. Full details of all courses and consultancy services can be obtained from:

The Conference Secretary,
Statistics for Industry (UK) Ltd,
14 Kirkgate,
Knaresborough,
North Yorkshire HG5 8AD.

Tel: (0423) 865955

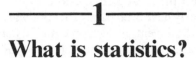

1
What is statistics?

'Statistics' is a word with many meanings. We are doubtful that a simple definition can be found which will be entirely satisfactory, even in such a specialized text. Let us therefore explore several popular definitions of statistics before selecting one for further consideration. These definitions will overlap to some extent but each emphasizes a point of some importance.

Definition 1: 'Statistics is a branch of mathematics'.
Definition 2: 'Statistics is a set of techniques which can be used to prove almost anything'.
Definition 3: 'Statistics is an extremely dull subject and its application involves very tedious calculations'.
Definition 4: 'Statistics is a body of knowledge which can be of use to anyone who has taken a sample'.

The first definition contains some truth. Many mathematicians study certain aspects of statistics and in doing so consider that they are exploring just another branch of mathematics. It is also true that a statistician needs a good grounding in mathematics especially if he or she is to engage in basic research. For the *applied statistician* however, an ability to communicate with his client (e.g. the analytical chemist) is just as important as mathematical expertise whilst the client for his part may be able to avoid *all* contact with mathematics. You may be very sceptical of this last statement. Perhaps you are a chemist who does not have access to an experienced applied statistician. Perhaps you have attempted to teach yourself statistics from a book which was very mathematical. Perhaps you were taught statistics alongside chemistry without being shown the connections between the two subjects. We assert nonetheless that the scientist or technologist can make use of statistics even if he has very little knowledge of mathematics. It is certainly true that you will find very little mathematics in this book.

The second definition also contains some truth. Clearly it is possible to deceive the unwary by presenting only a selected part of a set of data, or by reporting only those conclusions which support one's prejudices. It is also possible to arrive at

1

invalid conclusions by using statistical techniques inappropriately. Because of this possibility we include in later chapters a discussion of the *assumptions* underlying the recommended statistical techniques. Armed with a knowledge of these assumptions the scientist can safely make use of statistics and will be in a strong position to detect the invalid conclusions of others.

The third definition may have been partly true of some applications of statistics many years ago. With the advent of electronic digital computers, and especially with the recent proliferation of microcomputers, the drudgery of calculation has been eliminated. As for dullness of statistics, the reader must judge for himself, but the authors will be very disappointed if they fail to communicate the excitement of adapting statistical techniques to the particular problems of the analytical chemist.

The last of the four definitions is more useful as a foundation for introducing the basic ideas of applied statistics. Clearly the analytical chemist is well aware of the need to take *samples*. If however he is prepared to view the sampling process through the eyes of the statistician then the statistical techniques may seem more reasonable. We will explore two situations in which an analytical chemist has taken a sample. The first is not in a laboratory environment but it will serve to illustrate an important point.

EXAMPLE 1.1

Circulation of *The Analyst* in 1980 was approximately 9000 copies per month. Many copies are read by more than one person, of course, and the editor of *The Analyst* would like to *estimate* the average readership per copy. From the computer file of subscribers he selects 100 addresses at random and despatches a questionnaire to each. All of the questionnaires are returned and the editor calculates the average readership of the 100 copies to be 2.18. Is it safe for the editor to claim that the average readership is more than 2 readers per copy?
■ ■ ■

This very simple example illustrates several important points which would not be revealed so clearly in a more complex situation. The sample consists of 100 copies of *The Analyst* and the sample average is 2.18 readers per copy. The editor wishes to make a statement about all 9000 copies and a statistician would refer to this larger group as the *population*.

> A sample is simply a small group taken from a larger group about which we wish to draw a conclusion. This larger group is known as the population.

Clearly the validity of the conclusion will depend upon *how* the sample is drawn from the population. In this example the editor selected the 100 addresses at *random*. He was using a method known as *random sampling* which ensures that each member of the population has the same chance of being included in the sample. In many situations it is not possible to take a random sample as we shall see later.

Concerning the relationship between the sample and the population, several points need to be stressed:

(a) If the editor were to repeat the exercise he would almost certainly get a different sample and a different value for the sample average.
(b) Since average readership varies from sample to sample, it is rather unlikely that the editor's sample average (2.18) will be exactly equal to the population average (i.e. the average readership of all 9000 copies).
(c) Presumably the population average will be *close to* the sample average, but can the editor be sure that the population average will be greater than 2 readers per copy? (This question will be answered in a later chapter.)

EXAMPLE 1.2
An analytical chemist wishing to evaluate a new method carries out a preliminary investigation in which he makes six replicate determinations of the copper content of a solution which is known to have a copper content of 60.0 p.p.m. Each determination requires the preparation of a 10 ml sample and the determinations are:

$$58.2 \qquad 61.0 \qquad 56.6 \qquad 61.5 \qquad 53.8 \qquad 56.9$$

Can the analyst draw any conclusions from this limited amount of data?

The analyst would have been delighted if all six determinations had been exactly equal to 60.0. He expects, however, to find error in any determination and is not surprised that there is variability amongst the results. He is a little disappointed that the average determination (58.00) is not closer to the true concentration of 60.0 p.p.m. but he suspects that the difference might have been less if he had taken a larger number of determinations.
■ ■ ■

Clearly the analyst is interested in the *bias* and *precision* of the new method but before we discuss these important concepts we will draw attention to several features of the sample and the population. (You will recall that we earlier defined a sample as being a small group taken from a larger group known as the population.)

(a) In this situation the chemist might say that he had prepared *six samples* whereas the statistician would say that there is *one sample* containing six observations. Clearly there is some scope for confusion with the chemist focusing on distinct quantities of material and the statistician concentrating on sets of numbers. Throughout this book both meanings of the word 'sample' will be used but we hope that the meaning will be obvious from the context.

(b) Perhaps the benefit to be gained by taking a statistician's view of the sample will be clearer when we talk about the population. If the sample consists of six determinations then the population must also contain determinations, but more than six of them. We could define the population as being *all* determinations that the analyst might have made on the solution using the new method. If we accept this definition then the population is infinitely large and, furthermore, only six members of the population actually exist in a material sense. With such a population how are we to ensure that every member has the same chance of being included in the sample? It would appear to be impossible.

(c) Perhaps you would prefer not to talk about the population. It is intuitively obvious, however, that we are not likely to draw a valid conclusion unless our sample is representative of the population. Furthermore many statistical techniques do refer to a population and have an underlying assumption that the sample has been selected at random from this population.

(d) The analyst is more interested in the precision of the test method than in the hypothetical population put forward by the statistician. We will see in a later chapter how an estimate of precision can be obtained by measuring the variability in repeat determinations. We will also discuss techniques for comparing the precision of two test methods. All such procedures are, however, based on certain assumptions about the population from which the sample was taken.

(e) The analyst is also interested in the possibility that the test method might be biased. The average of the six determinations (58.00 p.p.m.) is certainly not equal to the true concentration (60.0 p.p.m.) but this does not prove that the new method is biased. Perhaps the sample average would have been closer to 60.0 p.p.m. if a larger number of determinations had been made. Is it possible that the sample average would have been *equal* to 60.0 p.p.m. if *all* possible determinations had been made on the solution? Clearly we can never answer such questions with absolute certainty but we will, in a later chapter, explore a technique which helps us to decide beyond reasonable doubt whether a test method is biased in any particular situation.

2

Describing a set of data

2.1 Introduction

Before we can explore the many areas in which the practising analytical chemist makes use of statistics, we must first examine some simple techniques which can be used to summarize a set of data. If we have two or more determinations we will naturally be concerned with the scatter of these measurements around some average value. To convey an impression of this scatter, to ourselves or others, a diagram can be very helpful and we will make frequent use of simple graphical techniques throughout this book.

Scatter and average can also be expressed in quantitative terms if we are prepared to carry out certain calculations on the data. These calculations are greatly facilitated by the use of a modern pocket calculator, which reduces the drudgery and increases our confidence in the results.

We will also introduce in this chapter the *normal distribution curve* which has proved useful in analytical chemistry for describing the scatter of errors. The use of this curve will always be based on an assumption that it is applicable to the situation under investigation.

2.2 Describing a small set of data

Let us take another look at the copper content determinations from the previous chapter. You will recall that the analytical chemist had made six repeat determinations on a solution which was known to have a concentration of 60.0 p.p.m. The results were:

$$58.2 \quad 61.0 \quad 56.6 \quad 61.5 \quad 53.8 \quad 56.9$$

Even with such a small set of data a graphical representation can be useful. A suitable graph can be obtained very easily if we represent each determination by a point (or a blob) on a line. The *blob chart* (Fig. 2.1) is a plot of the six determinations.

Marked on the blob chart are the true concentration (60.00) and the mean determination (58.00) *Mean* is a word very frequently used in statistics as an

5

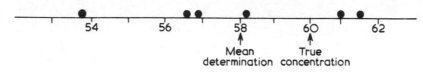

Figure 2.1   Six determinations of copper content – a blob chart

alternative to *average*. The mean determination is calculated by adding up the six determinations and then dividing the total by six. A formula for this operation is:

$$\text{Sample mean } (\bar{x}) = \sum x/n$$

In this formula $\sum$ is a Greek capital letter *sigma* and $\sum x$ represents the sum of the determinations, whilst n represents the number of determinations. The symbol $\bar{x}$ (pronounced x bar) is often used to represent a sample mean (if you are not familiar with sigma notation see Appendix 1. The symbols and formulae used are given in Appendix 2). It may seem pedantic to introduce a formula with Greek letters to describe the calculation of a simple average but this sigma notation will prove useful when we meet more complex calculations.

As an alternative to calculating the mean we could put the determinations into ascending order and select the middle one. This would be known as the *median* determination. With an even number of determinations we find that there isn't a middle *one* and we take the average of the middle *two*. In ascending order the six determinations are:

53.8   56.6   <u>56.9</u>   <u>58.2</u>   61.0   61.5

The median determination is $(56.9 + 58.2)/2$ which is 57.55 p.p.m.

Let us return to the blob chart in Fig. 2.1. It is probably true that the main value of such a diagram is that it can give an impression of the scatter or spread or variability of the determinations. The variability of a set of repeat determinations gives an indication of the precision (or lack of precision) of the test method. It is useful, therefore, to be able to quantify the spread of a set of measurements. This can be achieved by subtracting the smallest measurement from the largest to get the *range* of the sample.

$$\text{Sample range} = \text{highest value} - \text{lowest value}$$

For our six determinations the sample range is 61.5 minus 53.8 which is 7.7. If the new test method were less precise the six determinations would probably have been more widely spread and the sample range would then have been greater than 7.7. If all six determinations had been identical then the sample range would have been zero.

Though the sample range is very easily calculated it is not a very reliable measure of spread because it is over-dependent on the two most extreme values in the sample. A much more useful measure of spread is the sample standard deviation or its close relative the sample variance.

Sample variance $= \sum(x - \bar{x})^2/(n-1)$ OR $(\sum x^2 - n\bar{x}^2)/n - 1)$

Sample standard deviation $= \sqrt{}$(sample variance)

Use of these formulae will be illustrated by calculating the variance, and then the standard deviation, of the six determinations which have been listed in the left hand column of Table 2.1.

Table 2.1 Calculation of the variance of six determinations

	Determination x	Deviation from mean $(x-\bar{x})$	Squared deviation $(x-\bar{x})^2$	
	58.2	0.2	0.04	
	61.0	3.0	9.00	
	56.6	−1.4	1.96	
	61.5	3.5	12.25	
	53.8	−4.2	17.64	
	56.9	−1.1	1.21	
Total	348.0	0.0	42.10	$= \sum(x-\bar{x})^2$
Mean	58.00			

$$\text{Sample variance} = \sum(x - \bar{x})^2/(n-1)$$
$$= 42.10/5$$
$$= 8.42$$
$$\text{Sample standard deviation} = \sqrt{8.42}$$
$$= 2.902$$

In Table 2.1 the mean determination (58.00) has been subtracted from each individual determination to obtain the *deviations from the mean* in the second column. Note that some of the deviations are positive and some are negative whilst the column adds up to zero. Squaring the deviations gives the numbers in the third column and the total of this column (42.10) is known as the *sum of squares*. To get the sample variance we divide the sum of squares by $(n-1)$. This divisor is known as the *degrees of freedom* and we say that the sum of squares has $(n-1)$ degrees of freedom.

> Variance = (sum of squares)/degrees of freedom)

The question is often asked 'Why do we divide by $(n-1)$ and not by n?' In other words 'Why has the sum of squares got $(n-1)$ degrees of freedom?' Three points need to be stressed in answer to this question:

(a) *Occasionally* one might wish to divide by n when calculating a sample variance or standard deviation. Many pocket calculators will calculate the standard deviation very rapidly offering a choice of either n or $(n-1)$ as a divisor. The buttons are usually labelled σ_n and σ_{n-1}.

(b) To avoid confusion we will *always* divide by $(n-1)$ throughout this book when calculating a sample standard deviation.

(c) As we will see later, the purpose of calculating a sample standard deviation is almost invariably to estimate a population standard deviation. It can be shown mathematically (or by repeated experiments) that the use of n as a divisor will give a sample standard deviation which tends to *underestimate* the population standard deviation, whereas the use of $(n-1)$ gives what is known as an 'unbiased estimator'.

Both the variance and the standard deviation are measures of spread. Had the determinations been more widely scattered then both the standard deviation and the variance would have had larger values. Had all six determinations been equal then both measures of spread would have been equal to zero. When, you might wonder, would we use a standard deviation rather than a variance? An important point to note is that the sample standard deviation, like the sample mean, is expressed in the *same units* as the original data. If, for example, we were interested in the weights of analytical chemists and we recorded the weights of a sample of ten such chemists in pounds, then the mean weight would be in pounds and the standard deviation would also be in pounds, whilst the variance would be in pounds squared. The standard deviation is, therefore, of more direct use to the chemist, but variances will arise from time to time throughout this book and we will even use the sum of squares as a measure of variability in certain circumstances.

Quite often it is convenient to express a standard deviation as a percentage of a known value or of an average value. If we express the sample standard deviation as a percentage of the sample mean we obtain a dimensionless measure of spread known as the coefficient of variation (CV) or relative standard deviation.

> Coefficient of variation = (standard deviation/mean) × 100

Thus the coefficient of variation of the six determinations of copper content is
$(2.902/58.00) \times 100$ which is equal to 5.00%. For some purposes the coefficient of
variation may be more useful than the standard deviation as we shall see later.

2.3   Describing a large set of data

In addition to the six determinations we have just considered, the analytical
chemist has *many* determinations of the copper content of his standard solution
obtained by his *old* method. This particular solution has been used for quality
control purposes over a period of weeks and the sixty most recent determinations
are:

61.0	65.4	60.0	59.2	57.0	62.5	57.7	56.2	62.9	62.5
56.5	60.2	58.2	56.5	64.7	54.5	60.5	59.5	61.6	60.8
58.7	54.4	62.2	59.0	60.3	60.8	59.5	60.0	61.8	63.8
64.5	66.3	61.1	59.7	57.4	61.2	60.9	58.2	63.0	59.5
56.0	59.4	60.2	62.9	60.5	60.8	61.5	58.5	58.9	60.5
61.2	57.8	63.4	58.9	61.5	62.3	59.8	61.7	64.0	62.7

With this larger set of data it is difficult to get a feel for the spread of the
determinations and the need to summarize the sixty determinations is very clear.
We could handle this data from the old method in exactly the same way that we
treated the smaller set of data earlier, i.e. calculate the sample mean and the
sample standard deviation then draw a blob chart. With such a large set of data,
calculation of the standard deviation is rather tedious unless a suitable calculator
is available. Using such a calculator we find that the sample mean is equal to 60.37
and the sample standard deviation is equal to 2.541. (It is usual to round off the
mean to *one* decimal place beyond that used in recording the data and to round
off the standard deviation to *two* extra decimal places.)

As we can see in Fig. 2.2 the simple blob chart is not a very satisfactory
representation of this data, though it does give an indication that many of the

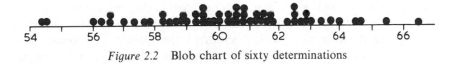

Figure 2.2   Blob chart of sixty determinations

sixty determinations are in the centre of the spread, from 58 to 63 say. With such a
large set of data there is a need to *group* the determinations before attempting a
pictorial representation. This grouping has been carried out in Table 2.2.

Table 2.2 is known as a *frequency distribution*. It indicates how the determinations
are spread or distributed by telling us how many determinations fall into each of
the thirteen groups. The numbers in the bottom row of the table are often referred
to as frequencies. We can see in Table 2.2 that two of the sixty determinations lie

Table 2.2   Frequency distribution of sixty determinations

Copper concentration (p.p.m.)	54.0 to 54.9	55.0 to 55.9	56.0 to 56.9	57.0 to 57.9	58.0 to 58.9	59.0 to 59.9	60.0 to 60.9	61.0 to 61.9	62.0 to 62.9	63.0 to 63.9	64.0 to 64.9	65.0 to 65.9	66.0 to 66.9	Total
No. of determinations	2	0	4	4	6	8	12	9	7	3	3	1	1	60

between 54.0 and 54.9 whilst there are none between 55.0 and 55.9 etc. It is very clear that a large percentage of the determinations lie in the middle five groups extending from 58.0 to 62.9.

It is often useful to have the frequencies of a frequency distribution expressed as percentages of the total frequency. If the numbers in the bottom row of Table 2.2 are divided by 60 and multiplied by 100 we get the percentage frequencies in Table 2.3.

Table 2.3   Percentage frequency distribution of sixty determinations

Copper concentration (p.p.m.)	54.0 to 54.9	55.0 to 55.9	56.0 to 56.9	57.0 to 57.9	58.0 to 58.9	59.0 to 59.9	60.0 to 60.9	61.0 to 61.9	62.0 to 62.9	63.0 to 63.9	64.0 to 64.9	65.0 to 65.9	66.0 to 66.9	Total
% of determinations	3.3	0.0	6.7	6.7	10.0	13.3	20.0	15.0	11.7	5.0	5.0	1.7	1.7	100%

The percentage frequency distribution in Table 2.3 tells us the percentage of the sixty determinations which fall into each group. The same information is presented in a graphical form in Fig. 2.3.

Each vertical bar in the histogram has a height which is proportional to the percentage of determinations in the group represented by the bar. We can see at a glance that the bulk of the distribution is bunched in the centre close to the true concentration of 60.0 p.p.m. Perhaps the most interesting parts of the histogram, however, are the two 'tails' which represent the highest and lowest determinations. These more extreme determinations will make an important contribution to our calculations when, in the next chapter, we discuss the difference that we are likely to find between two determinations made on the same sample. Unfortunately the histogram tells us very little about the largest or smallest determination we are likely to get since most of the information is bunched in the centre. We could of course obtain more determinations but however many we had we would always suspect that adding just one or two more determinations might make a substantial change to the shape of either tail of the distribution.

One way round this difficulty is to replace the histogram with a smooth curve

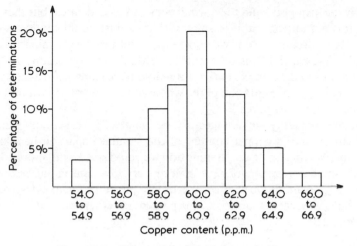

Figure 2.3 Histogram of sixty determinations

and one such curve which has proved very useful in analytical chemistry is known as the *normal distribution curve*.

2.4 The normal distribution

The histogram in Fig. 2.3 has been reproduced below together with a normal distribution curve. There are many normal curves. The one in Fig. 2.4 has been carefully chosen so that it has the same mean (60.37) and the same standard deviation (2.541) as the sixty determinations represented by the histogram.

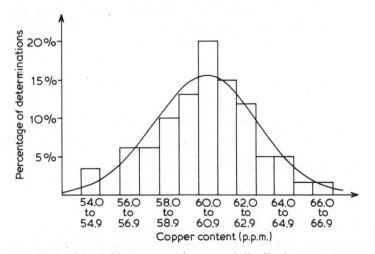

Figure 2.4 A histogram and a normal distribution curve

Clearly the histogram and the normal curve are not identical but they are very similar. It is not unreasonable to suggest that the histogram would get closer in shape to the normal curve if we took more and more determinations on the standard solution. The histogram is a faithful representation of the sixty determinations in the sample. Perhaps it would be reasonable to assume that the normal curve is a representation of the whole population of determinations from which the sample was taken.

There are certainly advantages to be gained by assuming that repeat determinations of the copper content of the standard solution have a normal distribution. Having made this assumption we could speak with more confidence about the small percentage of very high or very low determinations.

Suppose for example we wanted to estimate the percentage of determinations which would be greater than 65.0. This percentage is represented by the shaded area in Fig. 2.5.

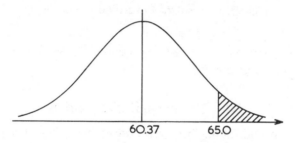

Figure 2.5 What percentage of determinations exceed 65.0?

The required percentage can be obtained from the *normal distribution table* (Table A at end of this book) but to use this table we must first *standardize* the value (65.0) in which we are interested. The deviation of this value from the mean must be expressed as a multiple of the standard deviation using the formula:

$$\text{Standardized value} = (\text{value} - \text{mean})/(\text{standard deviation})$$

$$\text{Standardized value} = (65.0 - 60.37)/2.541$$
$$= 1.822$$

This result is telling us that the value (65.0) is 1.822 standard deviations above the mean. Using a standardized value of 1.822 we can see from Table A that approximately 3.4% of determinations on the standard solution will exceed 65.0 p.p.m.

The normal distribution table can also be used 'in reverse'. Suppose for example we wished to know what value was likely to be exceeded by the highest 10% of determinations. Locating 10% in the right hand column of Table A we get

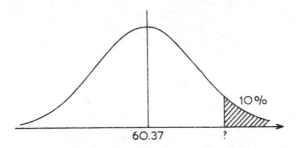

Figure 2.6   What value is exceeded by 10% of determinations?

a standardized value of 1.28, and we can now work backwards to find the unknown value:

$$1.28 = (\text{value} - 60.37)/2.541$$
$$\text{Value} = 60.37 + (1.282)(2.541)$$
$$= 63.628$$

We conclude that the top 10% of determinations made on the standard solution will exceed 63.63 p.p.m. approximately.

The use of Table A is very convenient but the validity of the conclusions we have drawn are dependent on the assumption that *repeat determinations on the standard solution have a normal distribution*. What grounds do we have for making this assumption?

The normal distribution curve has been studied by mathematicians for at least three hundred years and they have shown that, in situations where many small errors are influencing each measurement, the total errors in the measurement will have a normal distribution. It would be unwise to accept this assertion without supporting evidence. However, many scientists have found in a variety of situations that repeat measurements do appear to have a normal distribution. Furthermore there are many natural phenomena which appear to share this same distribution. For example the heights of adult males in Britain are reputed to have a normal distribution with a mean of 69 inches and a standard deviation of approximately 3 inches, whilst adult females have a similar distribution with a mean of 63 inches.

Many other physical features of living creatures are found to have a symmetrical bell-shaped distribution. In the field of mental abilities it is reputed that measured intelligence also has a normal distribution with males and females in Britain having the same mean (100) but different standard deviations.

For all of these normal distributions we can make use of Table A provided that we first convert the value (height, IQ, p.p.m., etc.) into a standardized value. Clearly *all* normal distributions have some common features and it is possible to make statements which refer to *all* variables which have a normal distribution, regardless of the mean or standard deviation.

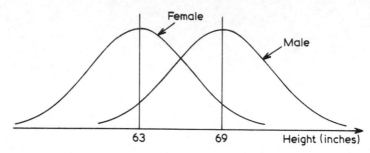

Figure 2.7   Distribution of height of adults in Britain

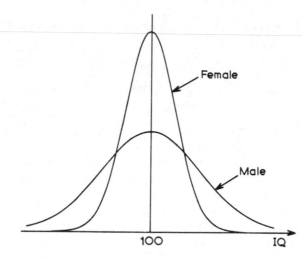

Figure 2.8   Distribution of intelligence of adults in Britain

Using μ to represent the mean and σ to represent the standard deviation we can say, for *any normal distribution*:

> 95% of values will lie in the range $\mu \pm 1.960\sigma$
>
> 99% of values will lie in the range $\mu \pm 2.575\sigma$

The numbers, 1.960 and 2.575, have been taken from Table A. They are the standardized values for 2.5% and 0.5% respectively.

2.5   Other distributions

It would be foolish to assume that the normal distribution is applicable to every

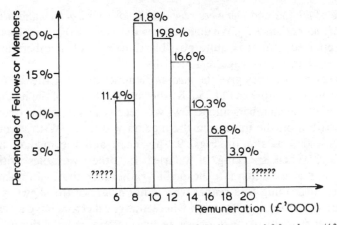

Figure 2.9 Annual remuneration of RSC Fellows and Members (1980)

set of measurements that the analytical chemist will take. There are other distributions which have been found useful in a variety of situations though we will return to the normal distribution repeatedly throughout this book. Many of the statistical techniques that we will use in later chapters are based on the assumption that errors of measurement will have a normal distribution. If this assumption is not valid then false conclusions may be drawn. Fortunately there are methods which can be used to check the assumption but we must always be watchful lest other distributions arise. We must be particularly wary of very *skewed* distributions such as the one in Fig. 2.9. The histogram in Fig. 2.9 is a graphical representation of the percentage frequency distribution in Table 2.4, which is based on a similar table in *Remuneration Sample Survey 1981* published by the Royal Society of Chemistry in March 1981.

Table 2.4   Annual remuneration of RSC Fellows and Members (1980)

Remuneration band	Under £6000	£6000 to £7999	£8000 to £9999	£10 000 to £11 999	£12 000 to £13 999	£14 000 to £15 999	£16 000 to £17 999	£18 000 to £19 999	£20 000 and over
% of Fellows and Members	2.4	11.4	21.8	19.8	16.6	10.3	6.8	3.9	6.9

You will notice that part of Table 2.4 is not represented in Fig. 2.9. No attempt has been made to represent in the histogram the 2.4% of members and fellows receiving less than £6000 or the 6.9% receiving £20 000 plus. Since we do not know the limits of these two groups we are unable to represent them accurately in Fig. 2.9. Despite these omissions the skewed shape of the distribution is very clear and is in contrast with the symmetrical shape of the histogram in Fig. 2.3.
An important point to note about such a skewed distribution is that the

percentage of fellows and members receiving below average remuneration will be considerably more than 50%. We do not know the average remuneration (because it is not reported) but it is quite possible that 70% of members and fellows received less than the average remuneration.

The survey report does give the *median* remuneration for the 4486 members and fellows in the sample as £11 350. By the very definition of the median we can say that half of the members and fellows will receive less than this figure. With a skewed distribution the mean and the median will differ. With a very skewed distribution, such as that in Fig. 2.9, the mean and the median may differ considerably. When speaking of salaries, or other variables with skewed distributions, it is usual to quote the median rather than the mean since the latter can be very misleading to a reader who is not aware of the skewness.

If you are operating in a situation where analytical errors have a very skewed distribution then many of the statistical techniques presented in this book may be of little use to you. Fortunately there is considerable evidence that repeat determinations within a laboratory have a symmetrical distribution which can be approximated by normal distribution. Furthermore there is evidence that variations from laboratory to laboratory also have a normal distribution (see Youden and Steiner, 1975).

2.6   Summary

In this chapter we have explored various ways of summarizing a set of data. We saw that a small set of data can be usefully represented by a simple blob chart. For a larger set of data the blob chart is difficult to draw and difficult to interpret. After grouping of the data into a frequency distribution, a histogram can be drawn.

A numerical summary of a set of data can be achieved by calculating the mean and standard deviation. These are easily obtained with the aid of a modern calculator. In later chapters we will make frequent use of the mean and standard deviation but we will repeatedly emphasize the value of simple diagrams in summarizing data.

If we are particularly interested in very high or very low values of the data we may prefer to use a normal distribution curve rather than a histogram. If we can reasonably assume that errors in repeat determinations have a normal distribution, then a normal curve may represent the *population* better than a histogram even though the latter faithfully represents the sample on which it was based.

In later chapters we will explore many practical situations in which we need to draw conclusions about a population using data taken from a sample. In drawing conclusions it will be very helpful to assume that errors have a normal distribution. Our immediate concern, in the next chapter, will be to classify errors and to discuss the difference we are likely to find between two repeat determinations.

Problems

(1) Three samples are submitted to a laboratory which carries out five determinations on each sample to give the following results:

Sample	A		B		C	
	6.1	6.3	36.5	37.3	241.5	244.7
Determinations	6.2	6.5	36.9	38.4	237.5	252.9
	5.9		35.4		242.9	

(a) Using an automatic routine on a calculator find the mean and standard deviation of each sample.
Calculate the coefficient of variation for each sample.
(b) Comment on the use of standard deviation and coefficient of variation as measures of precision for the analytical method.

(2) An autoanalyser is used to assay dioxamine on a regular basis. To control the operation of the analyser a determination is made at the beginning of each batch of samples on a standard of known concentration (200 mg/litre). If the autoanalyser finds the concentration to be outside 195.0 and 205.0 the analyser is declared to be malfunctioning.

It is known from previous experience that when the analyser is functioning correctly, determinations on the standard follow a normal distribution with a standard deviation of 2.0 mg/litre.

In what percentage of batches will the autoanalyser be declared to be malfunctioning:

(a) If it has zero bias and the correct precision.
(b) If it has a bias of +4.0 but the precision is correct.
(c) If it has a zero bias with the precision having deteriorated to give a standard deviation of 4.0.

The control procedure is declared to be unacceptable because of the low probability of detecting bias and therefore it is decided to change the limits.

(d) What should the limits be to give only a 10% chance of declaring that the analyser is working correctly when the bias is +4.0?
(e) Comment on the suitability of the control procedure for detecting either a bias of 4.0 or an increase in variability of 100%.

———3———
Errors and repeatability

3.1 Introduction

In the previous chapter we introduced the normal distribution and made use of the sample mean and the sample standard deviation. These statistical tools will prove useful when we discuss errors and the effect they have upon repeat determinations made in the same laboratory. You may recall from Chapter 2 that 95% of observations lie within 1.96 standard deviations of the mean if the observations have a normal distribution. We will make use of this fact when we estimate the difference that we are likely to find between two repeat determinations made in the same laboratory.

First we will discuss errors in a more general way and explore their effect upon the bias and precision of a test method.

3.2 Error, bias and precision

Every scientist is aware that any measurement he makes may be in error. In all branches of science it is acknowledged that measurement error can affect conclusions and that the reduction of error is therefore desirable. The methods adopted to reduce, or perhaps eliminate, error will depend upon the nature of the error and it is useful therefore to speak of three types:

(a) Random error.
(b) Systematic error.
(c) Gross error.

Random errors are irregular and unpredictable. They are present in all measurements and when repeat determinations have been made, random errors result in variability. Dr Smith has made six repeat determinations of the ammonia content of a delivery of ammonia solution. His results are:

   20.2%   19.9%   20.1%   20.4%   20.2%   20.4%

The determinations differ from each other because of random errors. If the test method did not introduce random error then the six determinations would be

18

identical (assuming that gross errors are absent). Clearly they are not identical and the magnitude of the random errors can be quantified by calculating the standard deviation of the determinations which is 0.190. We are unable to quantify the error in any *particular* determination unless we know the *true value* that we are attempting to measure. This is merely a statement of the obvious since we define the error as follows:

$$\text{Error} = \text{measured value} - \text{true value}$$

and

$$\text{Percentage error} = 100 \, (\text{measured value} - \text{true value})/\text{true value}$$

These formulae are of little practical use since only in certain contrived situations will the analytical chemist know the true value. In most situations he will only be made aware of random errors because of the variability he finds in repeat determinations.

Systematic errors, if present, affect a sequence of determinations equally. If determinations are made in batches, as is often the case, then a systematic error may result in every determination made within a particular batch being increased by a fixed amount whilst every determination in the following batch might be decreased by some other amount. If this were the case, we would speak of a '*fixed* systematic error' as distinct from a '*relative* systematic error'. The latter would result in all the determinations of a particular batch being increased (or decreased) by the same *percentage*. Throughout this chapter we will assume that any systematic error is of the fixed type but the very important relative systematic error will be discussed later.

Fixed systematic errors would not give rise to any extra variation within each batch but would cause extra variability between batches. A similar set of errors might arise if we had several operators or analysts carrying out repeat determinations. Mr Brown, Dr Jones and Miss Lee followed the example of Dr Smith and also made six determinations of the ammonia content of the same consignment of ammonia solution. The results presented by the four analysts are listed in Table 3.1 and represented by the blob charts in Fig. 3.1.

In each blob chart of Fig. 3.1, $\bar{x}$ represents the mean of the six determinations for the appropriate analyst. We see that the means for Smith, Jones and Lee are very similar (20.2, 20.0 and 20.2) whilst the mean for Brown (20.7) is rather different. We might conclude that each of Brown's six determinations contains a systematic error. Expressing the same conclusion more personally we might say that Brown was *biased*.

In making such a sweeping statement we need to bear in mind that we have only a limited amount of information about Brown's performance. In fact, when

Table 3.1   Repeat determinations by four analysts

Analyst	Determinations of ammonia content						Mean	SD
Smith	20.2	19.9	20.1	20.4	20.2	20.4	20.20	0.190
Jones	19.9	20.2	19.5	20.4	20.6	19.4	20.00	0.486
Brown	20.6	20.5	20.7	20.6	20.8	21.0	20.70	0.179
Lee	20.1	19.9	20.2	19.9	21.1	20.0	20.20	0.456

declaring Brown to be biased, we are commenting on a whole population of determinations after examining a rather small sample. Furthermore we do not know the *true* concentration of ammonia in the concentration of ammonia in the solution though we may have assumed it to be between 20.0 and 20.2 because the means for Smith, Jones and Lee lie in this range. Suppose that the true concentration were 20.1, as indicated by the broken line in Fig. 3.1, then we must ask ourselves 'Are we prepared to believe that Brown's mean determination

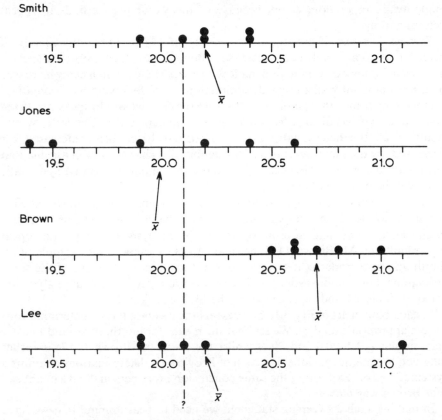

Figure 3.1   Repeat determinations by four analysts

would have been equal to 20.1 if he had made many more determinations on this delivery of ammonia solution?'.

In the next chapter we will explore an objective method for answering this question and we will take into account:

(a) The *difference* between Brown's sample mean (20.7) and the assumed population mean (20.1);
(b) The *sample size*, i.e. the number of determinations in the sample ($n = 6$);
(c) The *variability* that one expects to find amongst repeat determinations made by the same analyst.

Just as we distinguished between 'fixed' and 'relative' systematic errors we must also distinguish between 'fixed bias' and 'relative bias'. Brown would exhibit a fixed bias if all his determinations contained a fixed systematic error but he would be guilty of a relative bias if all his determinations were increased (or decreased) by a certain percentage. We will reserve the discussion of relative bias for a later chapter.

In Table 3.1 we have a standard deviation for each analyst (0.190, 0.486, 0.179 and 0.456). We note that two of these figures (0.486 for Jones and 0.456 for Lee) are much higher than the other two. Clearly the determinations made by Jones and Lee are more widely scattered than those made by Smith and Brown. Can we conclude that Jones and Lee are less precise? If Smith and Brown can achieve standard deviations as low as 0.190 and 0.179 we might wish to enquire 'Is it possible for Brown and/or Lee to achieve a similar standard?'.

In a later chapter we will explore a formal procedure that will help us to answer such questions. For the moment let us examine the spread of determinations in the blob charts in Fig. 3.1. We see that the pictures representing Smith and Brown are very similar except for the difference in means, noted earlier. Determinations made by Jones and Lee are spread across a much larger range, which is in keeping with their greater standard deviations. There is an important difference however between the two blob charts. Whilst in Jones's case the blobs are equally spread throughout the range, in Miss Lee's blob chart we find a tight cluster of five determinations together with an 'outlier' or 'flyer'. This 'rogue value' (21.1) does not appear to belong with the other five determinations. Perhaps it has been wrongly recorded or perhaps there was a chemical or procedural incident which would suggest that this particular determination does not belong to Miss Lee's normal population of repeat determinations.

If we were to reject the outlying determination (21.1) then Miss Lee's blob chart would have a spread similar to that of Smith or Brown and her standard deviation would fall to 0.130. Clearly a decision to reject a determination should not be taken lightly and we will discuss the 'rejection of outliers' more fully in a later chapter. On the other hand it is very unwise to rigidly process a set of data when there is strong evidence that one determination contains a *gross error*. Gross errors differ from random or systematic errors in that they are rare occurrences which do not fit into the usual pattern of errors associated with a particular situation. It is possible to pretend that gross errors do not occur. Perhaps they shouldn't but unfortunately they do.

To summarize the performance of the four analysts we might say that Smith and Brown have demonstrated high precision whilst Jones has not. Miss Lee could also be rated as precise provided we reject her outlying determination (21.1). Brown appears to be biased if we assume that the true ammonia content is 20.1%. Of the four analysts only Smith could be said to be *accurate* for he alone has presented clear evidence of high precision without bias; Miss Lee might also be declared accurate if reasons can be found to support the rejection of her 'rogue' determination.

3.3   Propagation of errors

A standardized test method or analytical procedure often involves more than one measurement being made, the final determination then being calculated from these measurements. If each of the individual measurements is in error then these errors will work their way through the calculation to give error in the determination. By exploring the way that the measurement errors are propagated we may be able to decide which is contributing most to the final error of determination.

Chemists in Indichem Ltd often sample-by-volume ammonia solution in bulk storage in order to asses the *density* of the solution at room temperature. The determination of density involves making three measurements which are then combined in the formula:

$$D = (M_2 - M_1)/V$$

where D = density of sample (g/ml)
    M_2 = mass of sample and flask (g)
    M_1 = mass of flask (g)
    V = volume of sample (ml)

If the two weighings are subject to error and the filling of the flask with ammonia solution is also subject to error, how will these errors affect the accuracy with which we determine the density? We noted earlier that accuracy depends upon *bias and precision*. It may therefore be useful to explore separately how the final determination is affected by:

(a) Systematic errors in the measurements;
(b) Random errors in the measurements.

A further subdivision of the problem is suggested by the arithmetic operations in the formula for density calculation. The two operations are subtraction and division. These will be considered separately.

3.3.1 *Addition and subtraction*

The first step in the calculation of the density of the ammonia solution is to obtain the mass of the sample using:

$$M = M_2 - M_1$$

where M = measured mass of solution (g)
   M_2 = measured mass of solution and flask (g)
   M_1 = measured mass of flask (g)

Errors in M_2 and M_1 will affect the value of M. If the weighing process is biased then both M_2 and M_1 will be subjected to a systematic error. If the systematic error is the *same* at both weights then the effect of the bias will be eliminated by the subtraction. As the weight of the full flask is roughly three times that of the empty flask (approximately 75 g and 25 g with a 50 ml flask) we may not be so fortunate. Perhaps the weighing process is subject to a bias such that the systematic error is proportional to the weight being measured (i.e. a relative bias). If this is so, then the two errors will not cancel in the subtraction.

Methods of *correcting* for bias will be considered later. Let us for the moment assume that the weighing is not biased and consider how the measurement of the mass of solution is affected by the *precision* of the weighings. The random errors which affect all weighings will manifest themselves in the form of *variability* when we make repeat weighings of the same item. By concentrating on variability and speaking of standard deviations rather than errors, we can make use of several well-known formulae.

$$SD(x - y) = \sqrt{[SD(x)^2 + SD(y)^2]} \qquad (3.1)$$

Equation (3.1) tells us how the variability of individual measurements, x and y, are related to the variability of the differences $(x - y)$. It is only applicable if random errors in the two measurements are independent of each other. Translating Equation (3.1) into a form which is suitable for our weighing problem we get:

$$SD(M) = \sqrt{[SD(M_1)^2 + SD(M_2)^2]}$$

Two points should be noted concerning this formula:

(a) The standard deviations are squared before adding. We are therefore adding variances rather than standard deviations. The fact that 'we can add variances but not standard deviations' has important consequences which will emerge in due course.
(b) The standard deviation of 'weight of solution' (M) must be greater than either the standard deviation of empty weighings (M_1) or the standard deviation of full weighings (M_2).
(c) If empty weighings and full weighings have *equal* precision (i.e. $SD(M_1) = SD(M_2)$) then:

$$SD(M) = \sqrt{(\sigma^2 + \sigma^2)}$$
$$SD(M) = \sigma\sqrt{2}$$

where σ = standard deviation of weighing.

If, therefore, we were to repeatedly fill the flask and estimate the weight of solution each time then these estimates would be approximately 41% more variable than the individual weighings as $\sqrt{2}$ is approximately equal to 1.41.

The implications of Equation (3.1) may be clearer if we consider three possible levels of precision for the weighing of the empty flask and observe the change in precision of the weight of solution:

EXAMPLE 3.1
Assume that the precision of weighing is such that repeat weighings of either the empty flask or the full flask have a standard deviation of 3 mg.

$$SD(\text{weight of solution}) = \sqrt{(3^2 + 3^2)}$$
$$= 4.24 \text{ mg}$$

■ ■ ■

EXAMPLE 3.2
Assume that the precision of weighing is such that repeat weighings of the empty flask have a standard deviation of 1 mg, whilst repeat weighings of the full flask have a standard deviation of 3 mg.

$$SD(\text{weight of solution}) = \sqrt{(1^2 + 3^2)}$$
$$= 3.16 \text{ mg}$$

■ ■ ■

EXAMPLE 3.3
Assume that the precision of weighing is such that repeat weighings of the empty flask have a standard deviation of 0.3 mg, whilst repeat weighings of the full flask have a standard deviation of 3 mg.

$$SD(\text{weight of solution}) = \sqrt{(0.3^2 + 3^2)}$$
$$= 3.01 \text{ mg}$$

■ ■ ■

We can see that *any* improvement in the precision of empty weighing will improve the precision of estimating the weight of solution. However, the improvement from Example 3.1 (4.24 mg) to Example 3.2 (3.16 mg) is much greater than that from Example 3.2 to Example 3.3 (3.01 mg). Clearly there is much to be gained by improving *either* weighing if the two are equally precise, but there is little to be gained by improving the more precise of the two weighings if they are already very different. Speaking more generally we can say that:

> If the errors in two stages of a determination are additive (or subtractive) and independent then the precision of the determination can be most easily improved by increasing the precision of that stage which is least precise (i.e. most variable).

In practice an analytical determination is often the culmination of a multistage process. Between the initial sampling and the final calculation is a whole chain of events and several people may be involved. It is well to remember that the easiest way to strengthen a chain is to strengthen the weakest link; though the weak link must first be identified, of course.

3.3.2 Multiplication and division

Having calculated the mass of ammonia solution in our 50 ml flask we can complete the determination of density using:

$$D = M/V$$

where D = density of sample (g/ml)
M = mass of sample (g)
V = volume of sample (ml)

How is the precision of the density determinations related to the precision of weighing and the precision of flask filling? Is there a simple equation similar to equation (3.1) which will relate the standard deviations of D, M and V? There is an equation, but unfortunately it is rather complex. A more useful but only approximate relationship can be expressed in terms of coefficients of variation:

$$CV(x/y) \simeq \sqrt{[CV(x)^2 + CV(y)^2]} \qquad (3.2)$$

(You will recall that the coefficient of variation is obtained by expressing the standard deviation as a percentage of the mean.)

Translating equation (3.2) into the language of our density determinations we get:

$$CV(D) \simeq \sqrt{[CV(M)^2 + CV(V)^2]}$$

To make use of this equation we need to know the coefficients of variation for weighing and filling. Suppose our past experience tells us that individual weighings of either empty or full 50 ml flasks have a standard deviation of 3 mg. As we have seen earlier:

$$SD(M) = \sqrt{[SD(M_1)^2 + SD(M_2)^2]}$$
$$= \sqrt{(3^2 + 3^2)}$$
$$= 4.24 \text{ mg}$$

The weight of ammonia solution in a 50 ml flask will be approximately 46.3 g, which gives a coefficient of variation as follows:

$$CV(M) = (0.004\,24/46.3) \times 100$$
$$= 0.009\%$$

Suppose that our past experience with 50 ml flasks suggest that the standard deviation of volume in repeat fillings is approximately 0.1 ml.

$$CV(V) = (0.1/50) \times 100$$
$$= 0.2\%$$

We can now combine these two coefficients of variation as follows:

$$CV(D) \simeq \sqrt{[CV(M)^2 + CV(V)^2]}$$
$$= \sqrt{[(0.009)^2 + (0.2)^2]}$$
$$= 0.2002\%$$

Using the density determination ($D = 46.3/50 = 0.926$) as a mean, we can convert the coefficient of variation back into a standard deviation as follows:

$$SD = CV \times mean/100$$
$$= 0.2002 \times 0.926/100$$
$$= 0.001\,85 \text{ g/ml}$$

Several points should be noted concerning the above calculations:

(a) Propagation of errors is not so straightforward when the calculations involve division (or multiplication) as compared with subtraction (or addition).
(b) The coefficient of variation of density determinations (0.2002%) is dominated by the lack of precision in the flask filling (CV = 0.20%) and is almost unaffected by the variability of weighing (CV = 0.009%).
(c) Because of the effect noted in (b) it would be pointless to increase the precision of the weighings. If improved precision of density determination is required this must come from improved precision of the flask filling. On the other hand, if the standard deviation of density determinations (0.001 85 g/ml) is acceptable then some relaxation of the precision of the weighings can be tolerated.

Our discussion of the propagation of errors, though limited to a single example, has been rather protracted. To help put the whole process in perspective a summary of the main steps and the calculated results is given in Fig. 3.2. The mean values in Fig. 3.2 are only estimates, and rather bad estimates, since they are based on single observations. The standard deviations are also estimates but they are based on considerable past experience. Estimation of means and standard deviations is very important in analytical chemistry and will be discussed fully in later chapters.

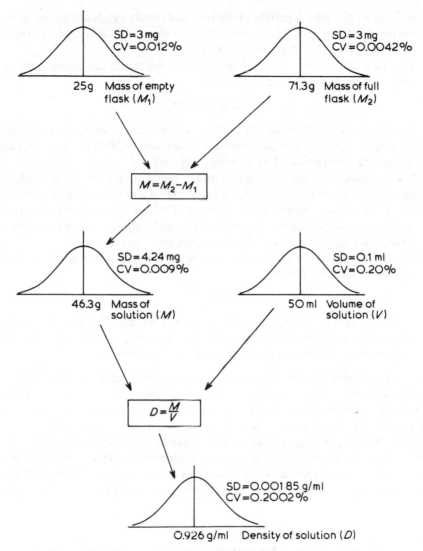

Figure 3.2 Propagation of errors in the determination of density

You will notice in Fig. 3.2 that each of the five variables has been given a normal distribution. The reasoning behind this allocation of distributions went as follows:

(a) It was *assumed* that repeat weighings of the empty flask and repeat weighings of the full flask would both have normal distributions. The means of the distributions differ, of course, but the standard deviations are equal, on the further assumption that the balance is equally precise whether weighing 25 g or 71 g. If your past experience tells you that repeat weighings do *not* have a normal distribution in a particular situation then it might be unwise to make the assumption that they do, of course.

(b) The mass of the solution is calculated by substituting the empty weighing (M_1) and the full weighing (M_2) into:

$$M = M_2 - M_1$$

It can be shown mathematically that M will also have a normal distribution if M_1 and M_2 have normal distributions. (The mean of M will be $(\mu_2 - \mu_1)$ and the standard deviation of M will be $\sqrt{(\sigma_1^2 + \sigma_2^2)}$.)

(c) It was *assumed* that the volume of liquid in repeat fillings of a 50 ml flask will have a normal distribution with a mean of 50 ml and a standard deviation of 0.1 ml. Is this a reasonable assumption? Perhaps some operators have a tendency to overfill rather than underfill, which might give rise to a skewed distribution and/or a mean which differed from 50 ml.

(d) The density is calculated by substituting the mass of solution (M) and the volume of solution (V) into:

$$D = M/V$$

It can be shown mathematically that D will have a distribution which is approximately normal if both M and V have normal distributions *and* their coefficients of variation are both small (i.e. less than 10%).

You may wonder why on earth we should be so concerned with the *shape* of the distributions of repeat measurements. It makes sense to worry anout the *mean* because this can indicate the presence of bias. It is equally sensible to be concerned with the standard deviation because this gives an indication of precision. You might imagine that the mean and standard deviation tell us *all* we need to know about a distribution of errors. Unfortunately this is not so for there are many useful techniques which are based on the assumption that errors have a normal distribution. This point arises when we consider the repeatability of a test method.

3.4   Repeatability

It is quite common for two (or more) determinations to be made on the same bulk material. For example, a customer *and* a supplier both assess the density of a consignment of ammonia solution. The price paid for the consignment depends in part upon the measured density, but which of the two measurements should be used when calculating the price? We could average the determinations and base the price on the mean, but this would be a reasonable procedure only if the two determinations are in reasonable agreement. Just how large a difference would we expect to find between two determinations when one is made in the customer's laboratory and the other is made on the supplier's premises?

This is a very important question which will be answered in a later chapter after we have explored ways of separating 'within-laboratory variation' from 'between-laboratory variation'. For the moment we will content ourselves with answering a much simpler question. 'How large a difference are we likely to find between two determinations made in the *same* laboratory by the same operator at approximately the same time?' In asking this question we are querying the *repeatability of the test method*. Two definitions of repeatability are given in British Standard 5532 *Statistics – Vocabulary and Symbols*:

> *Qualitative definition:* 'The closeness of agreement between successive results obtained from the same method on identical test material under the same conditions (same operator, same apparatus, same laboratory and short intervals of time.)'

> *Quantitative definition:* 'The value below which the absolute difference between two single test results obtained in the above conditions may be expected to lie with a specified probability. In the absence of other indication the probability is 95%.'

These definitions are referred to in many British Standards and are reproduced in BS 5497 *Precision of Test Methods* which is discussed in a later chapter. It will be clear from the definitions that repeatability is closely related to precision. A test method which lacks precision will have a high repeatability value and vice versa. To make clearer the quantitative definition of repeatability let us consider a specific example. Suppose that the repeatability of a test method is quoted as 0.213 then we would be 95% confident that two determinations made by this method under repeatability conditions would differ by less than 0.213. By 'repeatability conditions' we mean that the determinations were carried out by the same operator in the same laboratory with one determination made immediately after the other.

Obviously the repeatability of the test method (0.213) gives an indication of its precision under very homogeneous conditions. If two determinations are made under more heterogeneous conditions (e.g. in different laboratories) then we would not be surprised to find that they differed by more than 0.213.

To calculate the repeatability of a test method we need an estimate of the variability of repeat determinations. Earlier we estimated the standard deviation of repeat density determinations to be 0.001 85 g/ml. To obtain the repeatability of this test method we substitute $\sigma = 0.001\,85$ into the equation:

$$r = 1.96\sqrt{(2)}\sigma \tag{3.3}$$

where r = repeatability of test method
σ = standard deviation of repeat determinations

$$r = 1.96 \times 1.414 \times 0.001\,85$$
$$r = 0.005\,14 \text{ g/ml}$$

We would be 95% confident that two determinations of density made on the same consignment of ammonia solution in the same laboratory, by the same operator, would differ by less than 0.005 14 g/ml. If, on the other hand, two determinations made under these conditions were found to differ by more than 0.005 14 g/ml, then we would have grounds for suspicion that all was not well.

Equation (3.3) which was used to calculate the repeatability of the test method, is taken from BS 5497 *Precision of Test Methods*. An alternative equation will be presented in a later chapter but three points should be noted concerning equation (3.3):

(a) The $\sqrt{(2)}$ in the equation arises because we are talking about the difference between *two* determinations. If repeat determinations have a standard deviation equal to σ then the difference between pairs of repeat determinations will have a standard deviation equal to $\sqrt{(2)}\sigma$, as we have seen earlier.

(b) If we assume that repeat determinations have a normal distribution (with mean equal to the true density), then differences between pairs of repeat determinations will also have a normal distribution (with mean equal to zero). As we can see in Fig. 3.3, 95% of such differences will lie between $-1.96\sqrt{(2)}\sigma$ and $+1.96\sqrt{(2)}\sigma$.

(c) The σ in the equation represents the *population* standard deviation. In practice we would need to use a sample standard deviation and the equation would require modification to take account of the extra uncertainty. We will return to this point in the next chapter.

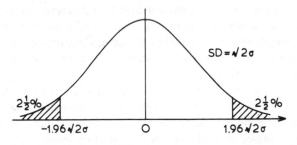

Figure 3.3   Distribution of differences between repeat determinations

In a later chapter we will calculate the *reproducibility* of a test method. This gives an indication of how large a difference we are likely to find when two determinations are made in *different laboratories*. Clearly repeatability and reproducibility are similar concepts but they apply to different situations. In fact they represent the two extremes one is likely to encounter. Repeatability refers to very homogeneous conditions with the same operator using the same equipment in the same laboratory. Reproducibility refers to very heterogeneous conditions with different operators using different equipment (perhaps even a different method) in different laboratories. Between these two extremes a whole spectrum of conditions exists and a very useful concept applicable throughout the range is the *least significant difference*. This tells us how large a difference we are likely to find between two determinations made under specified conditions. Thus the 'least significant difference under repeatability conditions' is simply what we have referred to earlier as 'repeatability'. We could, for example, refer to the 'least significant difference when a particular method is used by different operators in the same laboratory'. The numerical value of this least significant difference would lie between the repeatability and the reproducibility for this method.

3.5 Summary

In this chapter we have discussed two major concerns of the analytical chemist – bias and precision. In later chapters we will have a great deal more to say about both. The detection, estimation and correction of bias poses many statistical problems, whether the bias is fixed or relative. The estimation, comparison and improvement of precision also calls for the use of many statistical techniques.

Both bias and variability are the result of errors. We have classified errors as either random, systematic or gross errors and we have seen how random errors are propagated through the calculations which are often part of an analytical procedure. In a similar vein we estimated the size of difference one is likely to find between two determinations made in the same laboratory. A more sophisticated approach to repeatability will be taken when we later attempt to estimate the reproducibility of a test method.

The two formulae that we used when discussing the propagation of errors are reproduced below, together with two additional formulae that the reader might find useful:

$$SD(x+y) = \sqrt{[SD(x)^2 + SD(y)^2]}$$
$$CV(x \overset{\times}{\div} y) = \sqrt{[CV(x)^2 + CV(y)^2]}$$
$$\left. \begin{array}{l} SD(a+x) = SD(x) \\ SD(ax) = a\,SD(x) \end{array} \right\} \text{where } a \text{ is a constant}$$

Problems

(1) Clinical Chemistry Ltd wish to quote the repeatability for the determination of potassium in blood and consequently they devise an experiment in which the blood sample is analysed many times in a batch of determinations. The mean of the determination is 57.21 and the standard deviation is 1.225.

Calculate an estimate of the repeatability using the formula $r = 1.96\sqrt{(2)}\sigma$.

The repeatability is quoted to a customer who, on hearing of the design of the experiment, is very critical about several factors. These are:

(a) The operator might 'fiddle' the data when he has so many repeat determinations in a batch.

(b) Using the same batch for all determinations gives a too low indication of the variability.

(c) Only one operator was used for the repeatability and therefore the figure relates only to him and cannot be used generally.

(d) The repeatability figures relates to a mean determination of around 57 and could not be used with a considerably different mean level.

Comment on the validity of the customers' criticisms.

(2) British Standard 4359 gives standard analytical procedures for determining the specific surfaces of powders. Two of the procedures relate to the determination of the effective solid density (D) and the effective specific surface (VS). From these two parameters the effective mass specific surface (MS) is found by:

$$MS = VS/D$$

Both VS and D are subject to analytical errors which have been investigated for three different powders. The effect of these errors is to give the standard deviations below:

	Mean		Standard deviation	
Powder	VS	D	VS	D
A	200	20	40	2
B	400	60	40	6
C	600	100	40	10

For each powder:

(a) Calculate the coefficient of variation for determinations of both VS and D.

(b) Calculate the coefficient of variation for determinations of MS.

(c) Determine the standard deviation of MS.

(d) It is decided to increase the number of tests per sample for VS from one to four. What effect will this increase have on the standard deviation of MS? (*Note* the standard deviation of a sample mean of size 4 (called standard error) = standard deviation/$\sqrt{4}$.)

4

Fixed bias – detection estimation and correction

4.1 Introduction

In the previous chapter we distinguished between fixed bias and relative bias. The former results from each determination in a sequence or batch being increased (or decreased) by the same systematic error. A relative bias, on the other hand, results from each determination in a batch being shifted in the same direction by a constant percentage. Perhaps this is a rather narrow definition of bias. It is likely that the bias of a test method will have several components, each of which may be fixed or relative. For example there may exist a 'permanent bias' that is always present in addition to a bias that varies from batch to batch and a bias that varies from laboratory to laboratory.

In this chapter we will give further consideration to fixed bias. We will explore in some detail a systematic procedure which can help us to decide whether a set of repeat determinations contain evidence of a fixed bias. This procedure is a type of *significance test*. Other significance tests will be used in later chapters to compare the precision of two operators or to discriminate between fixed and relative bias.

We will also make use of simple formulae which enable us to estimate the magnitude of a fixed bias or to calculate the number of repeat determinations that are needed to detect a fixed bias of a specified magnitude.

4.2 Detection of a fixed bias

In Chapter 3 we posed the question 'Is Brown biased?' You will recall that Mr Brown and three other analysts had each made six repeat determinations of the ammonia content of a delivery of ammonia solution. It was believed that the true ammonia content was 20.1% but Brown's determinations had a mean of 20.70% and a standard deviation of 0.179%. Obviously 20.7 is *not* equal to 20.1 and no statistical technique can prove otherwise. We must bear in mind, however, that the 20.7 is a *sample mean* and we can only reasonably declare Brown to be biased if we are convinced that his *population mean* is not equal to 20.1.

The population in question could be defined as 'all determinations that Brown might have made on the consignment of ammonia solution.' We do not know the population mean. We never will know the population mean but we can follow a statistical procedure to decide whether or not Brown's population mean is equal to 20.1. This procedure is known as a one-sample *t*-test and is carried out in six steps as follows:

Step 1: *Null hypothesis* – Brown is not biased (i.e. Brown's population mean (μ) is equal to 20.1).

Step 2: *Alternative hypothesis* – Brown is biased (i.e. Brown's population mean (μ) is not equal to 20.1).

Step 3: *Test statistic*

$$\frac{|\bar{x} - \mu|}{s/\sqrt{n}} = \frac{20.7 - 20.1}{0.179/\sqrt{6}}$$
$$= 8.21$$

Step 4: *Critical values* – from the *t*-table (Table B) for a two-sided test with 5 degrees of freedom:

2.57 at the 5% significance level
4.03 at the 1% significance level

Step 5: *Decision* – We reject the null hypothesis.

Step 6: *Conclusion* – We conclude that Brown is biased.

The one-sample *t*-test above is the first of many significance tests that will be presented in this book. All these significance tests will share certain common features. They will all follow the same six-step procedure and they will all be carried out with the same purpose:

> The purpose of a significance test is to draw a conclusion about a population using data from a sample.

Let us go through the one-sample *t*-test step by step and note several points. If you have had no previous contact with significance tests you may find that some of these points are leading into deeper waters. Do not panic. Console yourself with the thought that *it is very easy to carry out a significance test* and your understanding will gradually develop.

(a) The null hypothesis is a simple statement about a population. We do not know whether the statement is true or false but it will be assumed to be true until proved otherwise.

(b) The alternative hypothesis is also a statement about the same population. As the alternative hypothesis is the 'opposite' of the null hypothesis (in the test we have carried out), you may wonder why the alternative is necessary. The need will become clear when we discuss one-sided and two-sided significance tests later.

(c) Note that neither hypothesis refers to the sample and that the number (20.1) referred to in both hypotheses does not come from the sample. Indeed, the two hypotheses could have been written down *before* the six determinations were carried out.

(d) Calculation of the test statistic condenses the data from the sample into one number. The formula used for this calculation depends upon the type of significance test being carried out.

For a one-sample *t*-test:

$$\text{test statistic} = \frac{|\bar{x} - \mu|}{s/\sqrt{n}}$$

where $\bar{x}$ = sample mean
  s = sample standard deviation
  n = sample size
  μ = hypothesized population mean

(When using this formula we take the difference between the two means and discard the minus sign which arises if $\bar{x}$ is less than μ).

(e) When calculating the test statistic in this particular test we took into account:

   (i) The deviation of the sample mean ($\bar{x}$) from the hypothesized population mean (μ).
   (ii) The variability between repeat determinations that we would expect to find when Brown is using this test method. This is quantified by the sample standard deviation.
   (iii) The number of repeat determinations made by Brown.

Obviously a larger value of test statistic would have resulted from either a larger difference between $\bar{x}$ and μ and/or a smaller standard deviation and/or a larger sample size. A large value of test statistic is an indication that Brown is biased.

In all the significance tests in this book a large test statistic implies that the null hypothesis is false.

(f) At the decision step we compare the calculated value of the test statistic with a critical value taken from the *t*-table. The critical value acts as a yardstick. If the test statistic is greater than the critical value we reject the null hypothesis. If the test statistic is less than (or equal to) the critical value we do not reject the null hypothesis.

(g) The critical values in any statistical table are calculated from formulae which are based on the assumption that the null hypothesis is true. Critical values can, therefore, be interpreted as values which are unlikely to be exceeded by the test statistic if the null hypothesis is true. For this one-sample t-test the 'critical value at the 5% significance level' is 2.57. There is less than 5% chance therefore that the test statistic would exceed 2.57 if Brown were not biased. Referring to the 1% significance level we could say that there is less than 1% chance that the test statistic would exceed 4.03 if Brown were not biased.

(h) As the calculated test statistic (8.21) is considerably larger than the quoted critical values it appears that either:

(i) Brown is not biased and a very unlikely event has occurred, or
(ii) Brown is biased.

We prefer to believe the latter, but obviously there is a small chance that we have made the wrong decision. Whenever we carry out a significance test we run a risk of making such an error. Unfortunately, we also run a risk of making another error, of a rather different nature. The two possible errors are highlighted in Table 4.1.

Table 4.1   Two possible errors in significance testing

| | Decision | |
| | Reject the null hypothesis and conclude that Brown is biased | Do not reject the null hypothesis and do not conclude that Brown is biased |
Reality		
Null hypothesis is false, i.e. Brown is biased	Correct decision has been made	Type II error
Null hypothesis is true, i.e. Brown is not biased	Type I error	Correct decision has been made

A type I error occurs if we conclude that Brown is biased when in fact he is not. On the other hand, a type II error occurs if we fail to conclude that Brown is biased though he actually is biased.

(i) The outcome of a significance test may well depend upon the significance level used. In Table B critical values are given for each of three significance levels; how does one decide which critical value to use? The choice of a significance level must depend upon the *risk* you are prepared to take of making a type I error. For some decisions a 10% significance level might be appropriate whereas for others 0.1% or even less would appear to be advisable. Choice of significance level depends upon the judgement of the scientist and is not left to a statistician. It is the duty of the statistician to advise, however, that reducing the significance level in order to reduce the risk of a type I error will also increase the risk of making a type II error. Fortunately it is possible to reduce both risks – by increasing the sample size. Commonsense would confirm that we would be in a better position to make judgement on Brown's bias if he had made a larger number of repeat determinations.

(j) Parallels can be drawn between the six-step procedure for significance testing and the procedure adopted in English criminal courts. We have assumed that the null hypothesis is true until proved otherwise, just as the defendant is assumed innocent until proved guilty. Absolute proof is not expected with either procedure; 'proof beyond reasonable doubt' is an acceptable compromise. The two possible verdicts 'guilty' and 'not guilty' correspond with the two possible decisions in a significance test 'reject the null hypothesis' and 'do not reject the null hypothesis'. We never accept the null hypothesis just as the courts never declare a defendant innocent. Two errors are possible with either procedure. A defendant may be found 'guilty' when in truth he is innocent or he may be found 'not guilty' though in fact he has actually committed the crime with which he has been charged. A jury which is over-concerned with reducing the chance of one error occurring may increase the risk of the other. In the criminal courts, as in significance testing, both risks will be reduced by an increase in the quantity and quality of evidence available.

Some of these points ((a)–(j) above) will no doubt appear perfectly reasonable and perhaps rather obvious whilst other points may be rather disturbing. Clearly significance testing is very important and it has many applications beyond the use of a one-sample t-test to detect bias. On the other hand, the mathematical basis of significance testing is rather complex and not within the scope of this book. You are therefore being invited to use a powerful technique that you do not fully understand. It may console you to learn that the majority of people who carry out significance tests are less than perfectly equipped for the task.

In the hope that familiarity with the technique will ease any anxiety you may feel, let us use the one-sample t-test to check for bias amongst the other analysts. You will recall that Miss Lee also produced six determinations of the ammonia concentration but we were reluctant to reach conclusions about her precision or bias because one of her determinations did not appear to fit in with the other five. We will reconsider Miss Lee when we discuss tests for outliers.

Dr Smith's six determinations had a mean of 20.20 and a standard deviation of 0.190. (See Table 3.1 in Chapter 3.) We noted that the deviation of Smith's mean from the true concentration (thought to be 20.1) was quite small and we concluded that Smith was not biased. What conclusion would we have drawn if we had used a one-sample t-test?

Null hypothesis –   Smith is not biased (i.e. Smith's population mean is equal to 20.1).

Alternative hypothesis – Smith is biased (i.e. Smith's population mean is equal to
20.1).

Test statistic $= \dfrac{|\bar{x} - \mu|}{s/\sqrt{n}} = \dfrac{20.20 - 20.1}{0.190/\sqrt{6}} = 1.29$

Critical values –   from the t-table with 5 degrees of freedom for a two-
sided test:

2.57 at the 5% significance level
4.03 at the 1% significance level
6.87 at the 0.1% significance level

Decision – We cannot reject the null hypothesis.
Conclusion – We are unable to conclude that Smith is biased.

Because the test statistic (1.29) is so small we cannot reject the null hypothesis at
any of the quoted significance levels. Note that we have *not* proved Dr Smith to be
unbiased. We have failed to prove that he *is* biased.

A one-sample t-test on the six determinations produced by Dr Jones gives a test
statistic of 0.50. Once again we are unable to conclude that the analyst is biased.

4.3   Estimating the magnitude of bias

The mean of Brown's six repeat determinations was 20.70 whilst the true
concentration was 20.1. The difference between these two figures is so large that
we have judged Brown to be biased.

Having proved beyond reasonable doubt that Brown's population mean is *not*
equal to 20.1 it is natural that we should now ask 'Just what *is* Brown's
population mean equal to?' In other words, having established that Brown is
biased we would like to know the magnitude of the bias.

The mean of Brown's determinations (20.70) is 0.6 above the true
concentration (20.1). It is tempting to conclude that Brown has a positive bias of
0.6, implying that all determinations Brown might have made on this
consignment of ammonia solution would contain a systematic error of $+0.6$. If
we did draw this conclusion we would almost certainly be wrong. We realize that,
if Brown were to carry out more determinations, his new mean would almost
certainly differ from 20.70 and it would seem wise that we should take into
account this variation from sample to sample when estimating the magnitude of
Brown's bias.

Perhaps it would be more prudent to quote a range of values rather than just
one number. We could for example claim that Brown has a bias which lies
between $+0.4$ and $+0.8$. Quoting an interval in this way would certainly improve
our chance of being right, but the interval would need to be infinitely wide for us
to be *certain* that it contained Brown's true bias. Statisticians advise that one
should quote an interval *and* a confidence level, the two together constituting
what is known as a *confidence interval*.

A confidence interval for a population mean is given by:

$$\bar{x} \pm ts/\sqrt{n}$$

where $\bar{x}$ is the sample mean;

s is the sample standard deviation;

n is the sample size;

t is taken from the two-sided t-table with $(n-1)$ degrees of freedom.

If we want to be 95% confident that the interval will contain the population mean we take the value of t from the 5% column in Table B and the result is known as a '95% confidence interval'. For a 99% confidence interval we use the 1% column.

Substituting $\bar{x} = 20.70$, $s = 0.179$, $n = 6$ and $t = 2.57$ into the formula gives a 95% confidence interval for Brown's population mean:

$$20.70 \pm 2.57(0.179)/\sqrt{6}$$
$$= 20.70 \pm 0.19$$
$$= 20.51\% \quad \text{to} \quad 20.89\%$$

We can be 95% confident that Brown's population mean lies between 20.51% and 20.89%. If Brown were to make a very large number of determinations on this consignment of ammonia solution, we could be 95% confident that his mean would lie between these two figures. The interval does *not* contain the true concentration (20.1%), which is hardly surprising as we have already established that there is a bias in Brown's performance. Obviously there will always be agreement between a one-sample t-test and a confidence interval for the population mean, provided that the same significance level is used. To test the null hypothesis '$\mu = 42$' say we could calculate a confidence interval for the population mean (μ) and reject the null hypothesis if the interval did not contain 42.

If we subtract the true concentration (20.1%) from the lower confidence limit (20.51%) and from the upper confidence limit (20.89%) we obtain a confidence interval for Brown's bias. The result of this subtraction tells us that we can be 95% confident that Brown's bias will lie between $+0.41$ and $+0.79$.

The quoting of an interval or a range of values appears eminently sensible to many analytical chemists. Use of the word 'confidence' however does not always receive immediate acceptance. Some analysts object to being told that they should be 95% confident. Others are not sure what '95% confidence' actually means. Perhaps it is a little strange that one person should tell another how confident he or she should feel.

The meaning that can be attached to a confidence interval will perhaps be more clear if we imagine that Brown had followed his first sample of six determinations with several other samples. Suppose that the results had been as shown in Table 4.2.

Table 4.2   Several samples of repeat determinations

Sample	Determinations	n	$\bar{x}$	s	t
A	20.6, 20.5, 20.7, 20.6, 20.8, 21.0	6	20.70	0.179	2.57
B	21.0, 20.5, 20.5, 20.0, 20.2, 20.8	6	20.60	0.276	2.57
C	20.6, 20.9, 21.1, 21.0	4	20.90	0.216	3.18
D	20.8, 20.6	2	20.70	0.141	12.71

Sample A in Table 4.2 contains the six determinations that we have examined previously and which gave a 95% confidence interval from 20.51% to 20.89%. Applying the confidence interval formula to the other three samples of repeat determinations gives the confidence intervals which are listed in Table 4.3.

Table 4.3   95% confidence intervals for Brown's true mean

Sample	Confidence interval	
A	20.70 ± 0.19	20.51 to 20.89
B	20.60 ± 0.29	20.31 to 20.89
C	20.90 ± 0.34	20.56 to 21.24
D	20.70 ± 1.26	19.44 to 21.96

We can see in Table 4.3 that each sample of determinations throws up a different confidence interval. If Brown had produced many many sets of determinations we would find that 95% of the resulting confidence intervals would contain Brown's population mean whilst the other 5% did not embrace this vital number. The four confidence intervals are also displayed graphically in Fig. 4.1.

We see that each confidence interval is centred on its sample mean whilst the

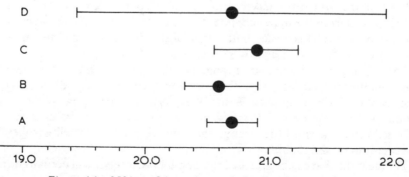

Figure 4.1   95% confidence intervals for Brown's true mean

width of each interval depends upon the sample standard deviation and the sample size. Note the alarming width of the 95% confidence interval given by sample D which is based on only two determinations. An estimate of the population standard deviation which has only 1 degree of freedom is extremely unreliable and the uncertainty in this estimate is reflected in the enormous value of t (12.71) which gives rise to the very wide confidence interval.

4.4   The sample size needed to estimate bias

It is intuitively obvious that, when carrying out a significance test or calculating a confidence interval, a large sample size is desirable. If we wish to check whether or not an operator is biased, then we stand a better chance of detecting any bias if the operator carries out a large number of repeat determinations. We did however get a very clear indication of Brown's bias with only *six* determinations. Are we to conclude therefore that six determinations will be adequate for any such investigation?

Unfortunately, it is not possible to give such a simple assurance. The sample size needed to estimate the magnitude of a bias will depend upon three factors:

(a) The limits within which we wish to estimate.
(b) The precision of the test method.
(c) The confidence we wish to have that the true bias will lie within the limits mentioned in (a).

On reflection factor (a) is not unreasonable. We are simply saying that a small sample may be adequate for obtaining a rough estimate of bias, but a large sample will be needed if we wish to estimate the bias with great precision. Factor (b) is also reasonable but rather disconcerting. If the test method had perfect precision there would be no random error, amongst which the systematic errors could hide, and any bias would be starkly revealed. In practice the test method is not perfect and we expect to find *variability* in repeat determinations. If this variability is excessive we may need a very large sample in order to estimate a bias with reasonable precision. Without doubt we will need an estimate of this random variability *before* we can calculate the size of sample that is needed in order to estimate the bias.

The sample size (n) needed to estimate a bias (by means of a confidence interval) to within $\pm c$ is given by:

$$n \simeq (ts/c)^2$$

where s is a previously obtained standard deviation with p degrees of freedom; and
t is taken from Table B with p degrees of freedom.

Suppose that Brown had been retrained in the use of the test method for the determination of ammonia concentration in ammonia solution. Naturally we would wish to determine whether or not the training had been effective. We could ask Brown to make several repeat determinations on a standard solution and use the results to estimate his bias. How many determinations should we ask him to make if we later wish to be 95% confident that we have estimated his bias to within ± 0.1?

To calculate the size of sample needed we will substitute $c = 0.1$ into the formula. To complete the calculation we will need a standard deviation and a value from the t-table. If we assume that Brown's precision will not have been changed by the retraining then we can use his previous standard deviation (0.179 from Table 3.1 of Chapter 3). As this standard deviation was calculated from six repeat determinations we use 5 degrees of freedom when taking the t-value from Table B. Substituting $t = 2.57$, $s = 0.179$ and $c = 0.1$ into the formula we get:

$$n \simeq (ts/c)^2$$
$$= (2.57 \times 0.179/0.1)^2$$
$$= 21.16$$

Thus we can be 95% confident that 22 repeat determinations will enable us to estimate Brown's bias to within ± 0.1. This is probably an over-estimate of the size of sample needed because the formula does not take into account that when Brown has carried out 22 determinations we will be able to use the new results to calculate a better estimate of his precision. Another point worth noting is that no guarantee is offered with this calculated sample size. Whether or not Brown's bias has been reduced by the retraining, we cannot be *certain* of estimating his true bias to within ± 0.1 from the results of the 22 determinations, but there is 95% chance that we will do so.

4.5   Correction of fixed bias

When Brown is engaged in the routine analysis of consignments of ammonia solution it will not be easy to tell whether or not his performance is biased. Each incoming consignment will have a different and *unknown* ammonia concentration. Furthermore Brown will make only *one* determination on each consignment. The results recorded by Brown will not therefore offer any scope for checking of bias or precision.

This problem can be overcome by using 'analytical quality control' procedures which will be described in a later chapter. These procedures depend upon the introduction of standard samples of known concentration at suitable points in a succession of routine unknown samples. Brown's determinations of concentration in these standards can then be used to check on his bias and precision.

An extension of this procedure would involve the introduction of a standard immediately after each unknown sample. Clearly this would double the cost of the analysis but it would allow us to remove any bias which might have affected the determinations made on the unknown samples. This correction could be carried out using the equation:

(corrected determination) = (determination on unknown sample)
− (determination on standard sample) + (true concentration of standard sample)

To study the effect of using this correction equation let us suppose that each of Brown's six determinations in Table 3.1 had been made in a separate batch and that each sample was followed immediately by a 10% standard. His results could have been as shown in Table 4.4.

Table 4.4 Determinations correction for fixed bias

| | Batches | | | | | | | |
	A	B	C	D	E	F	Mean	SD
Determinations on unknown samples	20.6	20.5	20.7	20.6	20.8	21.0	20.70	0.179
Determinations on standard samples	10.3	10.7	10.6	11.0	10.3	10.4	10.55	0.274
Corrected determinations	20.3	19.8	20.1	19.6	20.5	20.6	20.15	0.394

Each of the twelve determinations in Table 4.4 comprises a true value plus a random error plus a systematic error. The twelve random errors are independent of each other but the systematic errors are identical within any batch. To obtain the first of the corrected determinations we calculate:

$$20.6 - 10.3 + 10.0 = 20.3$$

Both the 20.6 and the 10.3 contain the same systematic error and therefore the effect of the bias is eliminated when we subtract. Both of these numbers contain random errors so that the corrected determination is contaminated by *two* random errors. For this reason the standard deviation of the corrected determinations (0.394) is greater than the standard deviations of the determinations on the standards (0.274) or the samples (0.179). The price we have paid for eliminating the bias is a reduction in precision.

(Our discussion of the propagation of errors in Chapter 3 would have led you to expect an increase in variability. Indeed you might have anticipated that the standard deviation of the corrected determinations would be equal to:

$$\sqrt{[(0.274)^2 + (0.179)^2]} = 0.327$$

rather than 0.394. The reason for this discrepancy will be explored in the next chapter.)

It should be noted that this method of bias correction is only appropriate if Brown has a *fixed bias*. Perhaps this is only likely to be the case if the sample and the standard have a very similar matrix. When the standard is more pure than the sample or the two matrices are thought to differ in some other respect then we would be wise to consider the existence of a *relative bias*. If it is thought that Brown has a *relative bias* then we should use an alternative correction formula which will be discussed in Chapter 5.

4.6   Repeatability again

In the previous chapter we calculated the repeatability of a test method using the formula:

$$r = 1.96\sqrt{(2)}\sigma$$

It was emphasized at the time that σ represents the standard deviation of a *population* of repeat determinations. In practice, of course, we will need to *estimate* the repeatability using a sample standard deviation and to take account of the extra uncertainty involved we must replace the 1.96 by a value from the *t*-table.

An estimate of the repeatability of a test method is given by:

$$r = t\sqrt{(2)}s$$

where s is the standard deviation of m repeat determinations under repeatability conditions;
  t is taken from the two-sided *t*-table with $(m-1)$ degrees of freedom.

Thus, if 8 repeat determinations were found to have a standard deviation equal to 0.284, we would calculate a repeatability estimate using $s = 0.284$ and $t_7 = 2.36$.

$$r = 2.36\sqrt{(2)}0.284$$
$$= 0.948$$

We would, therefore, expect two determinations made under repeatability conditions to differ by less than 0.948.

4.7   Summary

In this chapter we have introduced a procedure for carrying out a *significance test*. This procedure will be used again and again in later chapters. The basic principles that we have applied to the one-sample *t*-test will be used as a basis for a variety of significance tests throughout the book. In all these tests the purpose will be the same: 'To draw a conclusion about a population based on the data from a sample'.

In this chapter we have also taken a first step into *estimation*. We estimated the magnitude of a fixed bias for one particular operator using one particular method. When we come to discuss precision and calibration in later chapters, estimation will play a prominent part. In fact, estimation will be more important than significance testing as we calculate confidence intervals for the precision of a test method or for the concentration of X in an unknown sample.

Problems

(1) A new operator, Green, is given a sample containing a known concentration, 50 mg/litre, of chloride in water and instructed to make eight determinations. He obtains the following values:

$$49.4 \quad 49.8 \quad 50.8 \quad 49.3 \quad 51.3 \quad 50.0 \quad 50.8 \quad 51.8$$

(a) Is there evidence that Green is biased?
(b) Calculate a 95% confidence interval and hence state the maximum possible bias for Green.
(c) What sample size is needed to estimate Green's bias to within ±0.3 mg/litre?

(2) In order to remove Green's bias it is decided that in each batch there should be three determinations on a standard as well as three on each sample. The three determinations are then corrected using the formula:

$$\begin{pmatrix} \text{correction} \\ \text{determination} \end{pmatrix} = \begin{pmatrix} \text{determination} \\ \text{on unknown sample} \end{pmatrix}$$
$$- \begin{pmatrix} \text{determination} \\ \text{on standard sample} \end{pmatrix} + \begin{pmatrix} \text{true concentration} \\ \text{of standard sample} \end{pmatrix}$$

We can assume that determinations on both the standard and unknown samples have a standard deviation of 0.912 (the same as the previous problem).

(a) Calculate the width of a confidence interval for the true concentration of the unknown sample:

(i) Using the 3 uncorrected determinations;
(ii) Using the 3 corrected determinations.

Compare the widths of these intervals and hence decide what bias is required to make the correction method worthwhile.
(b) An alternative approach would be to calculate the mean of the three determinations on the unknown sample and calculate the mean of the

three determinations on the standard sample, then obtain *one* corrected
determination using:

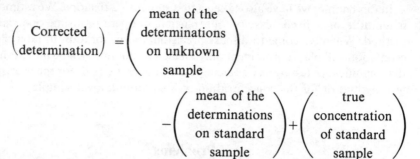

$$
\begin{pmatrix} \text{Corrected} \\ \text{determination} \end{pmatrix} = \begin{pmatrix} \text{mean of the} \\ \text{determinations} \\ \text{on unknown} \\ \text{sample} \end{pmatrix}
$$

$$
- \begin{pmatrix} \text{mean of the} \\ \text{determinations} \\ \text{on standard} \\ \text{sample} \end{pmatrix} + \begin{pmatrix} \text{true} \\ \text{concentration} \\ \text{of standard} \\ \text{sample} \end{pmatrix}
$$

Calculate the width of a confidence interval for the true concentration of
the unknown sample using the *one* corrected determination.

5

Relative bias – detection, estimation and correction

5.1 Introduction

In Chapter 3 we distinguished between fixed bias and relative bias. When an operator or a method exhibits a fixed bias all the determinations in a batch are affected by a systematic error of constant magnitude. With a relative bias on the other hand the systematic error is proportional to the concentration being measured so that a constant percentage error is introduced. In both cases the systematic error will be supplemented by a random error which will vary unpredictably from one determination to the next.

In Chapter 4 we concentrated upon fixed bias. We used a one-sample t-test to detect its presence, then we calculated a confidence interval to estimate its magnitude; finally we calculated the number of determinations needed to estimate a fixed bias within a certain tolerance.

In this chapter we will turn our attention to relative bias. To help us decide whether an operator has a fixed bias or a relative bias or both or neither, we will make use of a statistical technique known as *regression analysis*. This technique will again prove useful when we discuss calibration in later chapters.

5.2 Is the bias fixed or relative?

In the previous chapter we carried out a one-sample t-test which proved, beyond reasonable doubt, that Brown was biased. This conclusion applied only to this particular operator when using this particular test method on a solution with this particular concentration (approximately 20%). We later calculated a 95% confidence which indicated that Brown's bias was between $+0.41$ and $+0.79$. Once again this conclusion is only valid when referring to a 20% solution. We could not reasonably claim that Brown would have a similar bias if he were measuring the concentration of a 10% solution or a 5% solution. It is possible that the magnitude of Brown's bias depends upon the concentration being measured. Perhaps, with a 10% solution, we would find a bias which was only half that found with a 20% solution.

To investigate this possibility Mr Brown was asked to analyse a further five

47

prepared samples. These samples contained 5%, 10%, 15%, 20% and 25% ammonia, though Brown had no way of knowing these true concentrations. The results of this experiment are given in Table 5.1.

Table 5.1   Determination of ammonia concentration in five samples

						Mean	SD
True concentration x	5	10	15	20	25	15.0	7.906
Determination y	5.4	10.4	16.1	21.1	26.5	15.9	8.366
Deviation $(y-x)$	+0.4	+0.4	+1.1	+1.1	+1.5	0.9	0.485

There is certainly some evidence of bias in Table 5.1. All five of Brown's determinations have a positive deviation from the true concentration. We see however that the deviations are not constant. There is some indication that the deviations increase as the concentration increases. It is possible of course that this relationship might simply be due to random errors in the determinations. To help us reach a decision on this point the data have been plotted in Fig. 5.1.

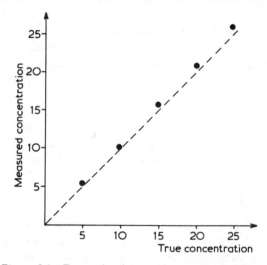

Figure 5.1   Determinations and true concentrations

The line in Fig. 5.1 is drawn at 45° to the axes. We would expect the points to be scattered around this line if, in fact, Brown were not exhibiting any bias, either fixed or relative. We see that all five points lie above the line. This is in keeping with the positive deviations in Table 5.1. Should we conclude that Brown has a fixed bias, a relative bias or neither? Perhaps Fig. 5.2 and 5.3 will help us to decide. They are graphs of hypothetical determinations in which rather exaggerated bias is present. In Figs. 5.2 and 5.3 the hypothetical data is not

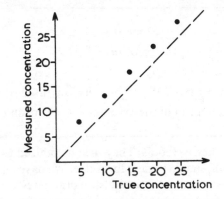

Figure 5.2   Evidence of a fixed bias $(y = 3.0 + x)$

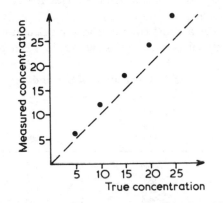

Figure 5.3   Evidence of a relative bias $(y = 1.2x)$

contaminated with random errors and the bias is, therefore, clearly visible. With the real life data in Fig. 5.1 we *do* have random errors and the random variation may obscure any bias which is present, be it fixed, relative or both. In an attempt to separate the random variation from any systematic variation we will fit a 'best straight line' to the data in Fig. 5.1 and then compare this line with 'best straight lines' fitted to Figs 5.2 and 5.3.

5.3   Fitting the 'best' straight line

Fitting the best straight line to a set of points simply involves finding values for a and b in the equation $y = a + bx$. We will use what is known as the 'method of least squares' which tells us that the slope of the line (b) and the intercept (a) are given by:

$$\text{Slope } (b) = Sxy/Sxx$$
$$\text{Intercept } (a) = \bar{y} - b\bar{x}$$
$$\text{where } Sxy = \sum(x - \bar{x})(y - \bar{y}) \text{ or } \sum xy - n\bar{x}\bar{y}$$
$$Sxx = \sum(x - \bar{x})^2 \text{ or } \sum x^2 - n\bar{x}^2 \text{ or } (n-1)(\text{SD of } x)^2$$
$$\bar{x} = \text{mean of the } x \text{ values} \qquad \bar{y} = \text{mean of the } y \text{ values}$$

'Sxx' and 'Sxy' are new symbols. They are introduced for the benefit of the reader because they will simplify many formulae. We have already met $\sum(x - \bar{x})^2$ in Chapter 2 when it was referred to as the 'sum of squares' rather than 'Sxx'. You will recall that we used the sum of squares when calculating a standard deviation. We can reverse this procedure and calculate the sum of squares (Sxx) from the standard deviation of the x values; which would seem wise if we have a calculator that has a rapid standard deviation facility.

Few calculators offer a short-cut method for calculating the 'sum of cross-products' Sxy. The calculation of Sxy is set out in Table 5.2.

Table 5.2 Calculation of Sxy and Sxx

True concentration x	Measured concentration y	$(x - \bar{x})$	$(y - \bar{y})$	$(x - \bar{x})(y - \bar{y})$	$(x - \bar{x})^2$
5	5.4	−10	−10.5	105.0	100
10	10.4	−5	−5.5	27.5	25
15	16.1	0	0.2	0.0	0
20	21.1	5	5.2	26.0	25
25	26.5	10	10.6	106.0	100
Total 75	79.5	0	0.0	Sxy = 264.5	Sxx = 250
Mean $\bar{x} = 15.0$	$\bar{y} = 15.9$	0.0	0.0		

$$\text{Slope } (b) = Sxy/Sxx$$
$$264.5/250$$
$$= 1.058$$
$$\text{Intercept } (a) = y - b\bar{x}$$
$$= 15.9 - 1.058(15.0)$$
$$= 0.03$$

The equation of the best straight line is $y = 0.03 + 1.058x$ and this line has been drawn in Fig. 5.4.

We can see that there is very little difference between the best line and the 45° line; but this small difference may be of great importance. In the application of

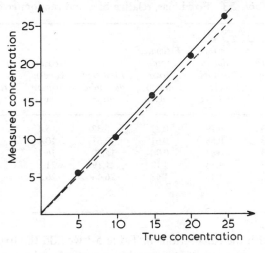

Figure 5.4   Best straight line

statistical techniques the analytical chemist is often looking for a very small difference between two very large numbers. In his statistical analysis, just as in his chemical analysis, he may be hunting a needle in a haystack. We will see in later chapters that great care must be taken with statistical calculations if the needle is to be found.

Despite the closeness of the two lines in Fig. 5.4 we must ask ourselves 'Does the difference between the lines suggest that we have a fixed bias (as in Fig. 5.2) or a relative bias (as in Fig. 5.3)?' We need to take into account that a fixed bias gives rise to a best straight line which has the same slope as the 45°–line but with a *different intercept* (as in Fig. 5.2) whilst a relative bias gives rise to a best line which passes through the origin but has a *different slope* to the 45° line (as in Fig. 5.3). Thus a fixed bias will result in *an intercept which is not equal to zero* (e.g. $a = 3.0$ in Fig. 5.2) and a relative bias will result in *a slope which is not equal to unity* (e.g. $b = 1.2$ in Fig. 5.3). The fitted equation $y = 0.03 + 1.058x$ would suggest therefore that Brown has a fixed bias of $+0.03$ *and* a relative bias of $+5.8\%$ (as 1.058 is 105.8%).

Using these two findings we can calculate the bias that we would expect Brown to have at any particular concentration of ammonia solution. We could then calculate what value of determination we would expect him to obtain at that concentration. These calculations have been carried out for each of the five concentrations used in the experiment and the results are tabulated in Table 5.3.

The fixed bias gives a constant error of $+0.03$ regardless of the true concentration. The relative bias gives an error which is directly proportional to the true concentration ranging from 0.29 for a 5% solution to 1.45 for a 25% solution. The relative bias certainly appears to exert a greater influence on the determinations than does the fixed bias.

Table 5.3   Fixed bias, relative bias and random error

True concen-tration (x)	Estimated error due to fixed bias (a)	Estimated error due to relative bias (0.058x)	Estimated total error due to bias	Predicted determin-ation	Measured concen-tration (y)	Residual (estimated random error)	Squared residual
5	0.03	0.29	0.32	5.32	5.4	0.08	0.0064
10	0.03	0.58	0.61	10.61	10.4	−0.21	0.0441
15	0.03	0.87	0.90	15.90	16.1	0.20	0.0400
20	0.03	1.16	1.19	21.19	21.1	−0.09	0.0081
25	0.03	1.45	1.48	26.48	26.5	0.02	0.0004
						0.00	0.099

Continuing with the calculations in Table 5.3 we add the fixed bias error and the relative bias error to obtain the total bias error. Further addition of the true concentration gives the predicted determination. This is the value of determination we would expect Brown to produce if he were to repeat the exercise. Each predicted determination is very close to the actual determination (or measured concentration) in the next column; the difference between the two is known as a *residual*. The residuals are estimates of the random errors in the determinations and they can be used to estimate the precision of the test method:

Residual = measured value − predicted value

You will note that some residuals are positive whilst others are negative and the sum of the residuals is equal to zero. The positive residuals correspond to the points in Fig. 5.4 which lie *above* the best straight line and the negative residuals correspond to those points which lie *below* the line. In fact the residuals could have been obtained by measuring the vertical distances of the points from the fitted line.

The sum of the residuals is 0.00. It can be shown that any line which passes through the centroid (the point at which $x = \bar{x}$ and $y = \bar{y}$) will give residuals which add up to zero. Our best straight line passes through the centroid where $x = 15$ and $y = 15.90$. The sum of the squared residuals (0.099) in the final column of Table 5.3 is known as the *residual sum of squares*. This sum of squares is a very useful measure of how well the line fits the points. If all five points lay on the fitted line then each of the residuals would be equal to zero and the residual sum of squares would be zero. It can be shown that the fitted equation $y = 0.03 + 1.058x$ represents a line which has *a smaller residual sum of squares than any other line would have* (with this set of data). In this sense, therefore, the fitted line is the *best* straight line and it is perfectly reasonable to refer to the method we have used as 'the method of least squares'. The residual sum of squares could have been calculated perhaps more quickly using:

$$\text{Residual sum of squares} = Syy - b^2(Sxx)$$

We have said earlier that the residuals are estimates of the random errors in Brown's determinations and that the residuals can be used to estimate the precision of the test method. To calculate such an estimate we must convert the residual sum of squares into a variance and then into a standard deviation. To achieve this we divide the sum of squares by its degrees of freedom then take the square root.

If a straight line has been fitted to a set of points then the residual sum of squares will have $(n-2)$ degrees of freedom, where n is the number of points. We lose two degrees of freedom because we have estimated both the slope and the intercept. Dividing the residual sum of squares by $(n-2)$ will give the residual variance and then taking the square root will give the residual standard deviation:

$$\text{Residual variance} = (\text{residual sum of squares})/(n-2)$$
$$= 0.099/3$$
$$= 0.033$$
$$\text{Residual standard deviation} = \sqrt{(\text{residual variance})}$$
$$= \sqrt{(0.033)}$$
$$= 0.1817$$

The residual standard deviation is an estimate of the variability we would expect to find if Brown made repeat determinations on the same sample, regardless of the true concentration. (We are assuming that Brown's precision does *not* depend on the concentration. This assumption will be discussed fully in the next chapter). 'Surely' you will say 'we already have such an estimate'. In Table 3.1 we have six determinations made by Brown on a consignment of ammonia solution which was thought to have a concentration of 20.1%. These six determinations had a mean of 20.70 and a standard deviation of 0.179. Though this latter figure is not exactly equal to the value of the residual standard deviation (0.1817) the difference between the two is so small that we can easily accept that the two experiments have produced estimates of the same population standard deviation. Both estimates are therefore indicative of Brown's precision.

We must take account of Brown's precision when we calculate confidence intervals for the true intercept and the true slope. If we were to repeat this investigation of Brown's bias we would almost certainly get slightly different determinations and, as a result, different values of intercept (a) and slope (b). This would serve to remind us that the five determinations in Table 3.1 are only a *sample* and that the calculated values of 'a' and 'b' are only estimates of the true slope and the true intercept.

A confidence interval for the true intercept is given by:

$$a \pm t(\text{ESD}) \sqrt{\left(\frac{1}{n} + \frac{\bar{x}^2}{Sxx}\right)}$$

where a = the calculated intercept

t is taken from the t-table with appropriate degrees of freedom

ESD = the error standard deviation

n = the number of points

Brown's error standard deviation (ESD) could be estimated in many ways. We could use the standard deviation of his repeat determinations (0.179) from Chapter 3 or we could use the residual standard deviation (RSD) that we have just calculated. We will use the RSD which has 3 degrees of freedom, but we will discuss this point further in the next chapter.

Using $a=0.03$, $t=3.18$ (3 degrees of freedom and 5% significance), RSD $=0.1817$, $n=5$, $\bar{x}=15.0$ and $Sxx=250$, we can calculate a 95% confidence interval for the true intercept:

$$0.03 \pm 3.18(0.1817) \sqrt{\left[\frac{1}{5} + \frac{(15.0)^2}{250}\right]}$$

$$= 0.03 \pm 3.18(0.1817)1.049$$

$$= 0.03 \pm 0.61$$

$$= -0.58 \quad \text{to} \quad +0.64$$

We can be 95% confident that the true intercept lies between -0.58 and $+0.64$. You will recall that the intercept is an indication of fixed bias; so we can be confident that Brown's fixed bias will lie between -0.58 and $+0.64$. As this interval includes zero *it is quite possible that Brown has no fixed bias at all.* Let us therefore focus our attention on his variable bias by calculating a confidence interval for the true slope:

A confidence interval for the true slope of a regression line is given by:

$$b \pm t(\text{ESD})/\sqrt{(Sxx)}$$

where b = calculated slope

t is taken from the t-table with appropriate degrees of freedom

ESD = error standard deviation

Using $b = 1.058$, $t = 3.18$ (3 degrees of freedom and 5% significance), $ESD = RSD = 0.1817$ and $Sxx = 250$ we can calculate a 95% confidence interval for the true slope:

$$1.058 \pm 3.18(0.1817)/\sqrt{250}$$
$$= 1.058 \pm 0.037$$
$$= 1.021 \quad \text{to} \quad 1.095$$

We can be 95% confident that the true slope lies between 1.021 and 1.095. Converting these two confidence limits into percentages gives 102.1% and 109.5% so we can be 95% confident that Brown's relative bias lies between a 'systematic increase of 2.1%' and a 'systematic increase of 9.5%'. As this range does *not* include a systematic error of 0.0%, we have proved beyond reasonable doubt that Brown has a relative bias.

We can use the regression equation $y = 0.03 + 1.058x$ to predict the determination that Brown would produce if he were asked to analyse ammonia solution which had a particular concentration. If, for example, Brown were given a 17% solution we would expect his determination of concentration to be equal to:

$$0.03 + 1.058(17) = 18.016\%$$

We realize that this prediction could prove to be incorrect for any particular sample of solution. In fact, if Brown were given many samples of 17% solution, we would expect his determinations to vary from sample to sample around an average of approximately 18.016%. With this in mind we can calculate two distinctly separate confidence intervals:

(a) A confidence interval for the mean determination that Brown would obtain with a *population of samples* having a specified concentration (X) is given by:

$$a + bX \pm t(ESD)\sqrt{\left[\frac{1}{n} + \frac{(X - \bar{x})^2}{Sxx}\right]}$$

(b) A confidence interval for the determination that Brown would obtain with a *particular sample* having a specified concentration (X) is given by:

$$a + bX \pm t(ESD)\sqrt{\left[1 + \frac{1}{n} + \frac{(X - \bar{x})^2}{Sxx}\right]}$$

Substituting into these formulae $X = 17$, $t = 3.18$ (5% significance, 3 degrees of freedom), $ESD = RSD = 0.1817$, $n = 5$, $\bar{x} = 15$ and $Sxx = 250$ we obtain the following confidence intervals:

95% confidence interval for *mean* determination on 17% solution

$$= 0.03 + 1.058(17) \pm 3.18(0.1817)\sqrt{\left[\frac{1}{5} + \frac{(17 - 15)^2}{250}\right]}$$
$$= 18.016 \pm 0.578\sqrt{(0.2 + 0.016)}$$
$$= 18.016 \pm 0.269$$
$$= 17.747 \quad \text{to} \quad 18.285$$

95% confidence for a *particular* determination on 17% solution

$$= 0.03 + 1.058(17) \pm 3.18(0.1817) \sqrt{\left[1 + \frac{1}{5} + \frac{(17-15)^2}{250} \right]}$$
$$= 18.016 \pm 0.578 \sqrt{(1 + 0.2 + 0.016)}$$
$$= 18.016 \pm 0.637$$
$$= 17.379 \quad \text{to} \quad 18.653$$

It is clear that we get a much wider confidence interval when we are estimating the result of one particular analysis than when we are estimating Brown's long-term mean determination. This would be true whatever concentration we had used. It should be noted, however, that *both* confidence intervals would have been wider if we had chosen a concentration (X) further from the mean value of x used in the experiment $(\bar{x} = 15)$.

At this point let us summarize the conclusions we can draw from the results of our regression analysis. From the confidence intervals that we have calculated it appears that Brown:

(a) has no significant fixed bias;
(b) has a relative bias between $+2.1\%$ and $+9.5\%$;
(c) would have a *mean* determination between 17.75 and 18.29 if he made repeat determinations on a 17% sample;
(d) would obtain a determination between 17.38 and 18.65 if he made *one* determination on a 17% sample.

One feature shared by all of these confidence intervals is that they are disappointingly wide. This width is a direct result of our attempting to do a great deal with very little data. From Brown's five determinations we have estimated:

(a) the intercept or fixed bias (thus losing one degree of freedom);
(b) the slope or relative bias (thus losing a second degree of freedom); and
(c) residual standard deviation or precision (with the remaining three degrees of freedom).

Perhaps the easiest way to reduce the width of our confidence intervals would be to obtain a better estimate of Brown's precision. We will discuss the estimation of precision in later chapters. Let us now turn our attention to the possibility that we may not have been using the best method to estimate Brown's relative bias.

5.4 Fitting the 'best' straight line through the origin

Having fitted the regression line $y = 0.03 + 1.058x$ we calculated a confidence interval for the true intercept and, as this interval included zero, we were unable to conclude that Brown had a fixed bias. We then went on to calculate a confidence interval for the true slope in order to estimate Brown's relative bias. It could be argued that we should have discontinued this analysis when we found that the true intercept might be equal to zero. If we are *certain* that Brown has no fixed bias, then the best way to estimate his relative bias is to fit the equation $y=bx$ which represents a line passing through the origin. Using the method of least squares the slope of this line is estimated by:

$$\text{Slope } (b)=\sum xy/\sum x^2$$

Table 5.4 Calculation of $\sum xy$ and $\sum x^2$

True concentration x	Determination y	xy	x^2
5	5.4	27.0	25
10	10.4	104.0	100
15	16.1	241.5	225
20	21.1	422.0	400
25	26.5	662.5	625
		$\sum xy=1457.0$	$\sum x^2=1375$

$$\begin{aligned}\text{Slope } (b) &=\sum xy/\sum x^2 \\ &=(1457.0)/(1375) \\ &=1.059\,636 \\ &\simeq 1.060\end{aligned}$$

The equation of the best straight line through the origin is $y=1.060x$. This equation indicates that Brown has a relative bias of $+6.0\%$, which differs little from the $+5.8\%$ estimated by the best straight line through the centroid. Note that the best line through the origin is unlikely to pass through the centroid and for this data it does not. If the new line were drawn on Fig. 5.4 it would be barely distinguishable from the existing line. Perhaps the best way to appreciate the difference between the two is by comparing the residuals in Table 5.5 with those in Table 5.3.

We see that the residuals in Table 5.5 do *not* add up to zero as they certainly would if the line passed through the centroid. We also see that the residual sum of squares (0.099 575) for the line through the origin is greater than the residual sum-of-squares (0.099) for the best line fitted earlier. Clearly the new line does not fit quite as well as the best line through the centroid. Nonetheless, the slope of the new line (1.060) gives us a better estimate of Brown's relative bias, provided that we are prepared to *assume* that he has no fixed bias.

Table 5.5 Relative bias and random error

True concentration (x)	Estimated error due to relative bias (0.05964 x)	Predicted determination	Measured concentration (y)	Residual	Squared residual
5	0.298	5.298	5.4	0.102	0.010404
10	0.596	10.596	10.4	−0.196	0.038416
15	0.895	15.895	16.1	0.205	0.042025
20	1.193	21.193	21.1	−0.093	0.008649
25	1.491	26.491	26.5	0.009	0.000081
				0.027	0.099575

The residual sum-of-squares could have been calculated from:

$$RSS = \sum y^2 - b^2 \sum x^2$$

which is very similar to the formula $(RSS = Syy - b^2 Sxx)$ used earlier with the best straight line through the centroid. To calculate the residual standard deviation we divide this sum-of-squares by its degrees of freedom then take the square root. Only *one* degree of freedom was lost whilst fitting the line through the origin as there was no intercept to estimate. It is appropriate therefore to use $(n-1)$ degrees of freedom.

$$\text{Residual standard deviation} = \sqrt{(0.099\,575/4)}$$
$$= 0.1578$$

The residual standard deviation (RSD) can be used as an estimate of the error standard deviation (ESD) in the calculation of various confidence intervals using formulae which are strikingly similar to those used earlier. For example, a confidence interval for the true slope of the line is given by:

$$b \pm t(\text{ESD})/\sqrt{x^2}$$
$$= 1.060 \pm 2.78(0.1578)/\sqrt{1375}$$
$$= 1.060 \pm 0.012$$
$$= 1.048 \quad \text{to} \quad 1.072$$

We can be 95% confident that the true slope lies between 1.048 and 1.060. We can be equally confident that Brown's relative bias lies between +4.8% and +7.2%. Note that this interval is narrower than that which was based on the line through the centroid (+2.1% to +9.5%). In calculating the new interval we used a *t*-value that was smaller, a residual standard deviation that was smaller, and $\sum x^2$ instead of Sxx. It is the latter change which has made the greatest difference.

There are certainly advantages to be gained by fitting a straight line through the origin rather than fitting a line through the centroid. There are also certain dangers, which will be discussed fully when we use regression equations for calibration purposes in later chapters.

5.5   Correction of relative bias

In the previous chapter we discussed how standard samples could be used to correct for fixed bias in Brown's determination of the concentration of ammonia in unknown samples. We used the correction formula:

(corrected determination) = (determination on unknown sample)

− (determination on standard sample) + (true concentration of standard sample)

This formula will only be appropriate if Brown has a fixed bias and no relative bias. We have now proved beyond reasonable doubt that Brown has a relative bias but no fixed bias; it would therefore be appropriate to use the correction formula below:

(corrected determination) = (determination on unknown sample)

÷ (determination on standard sample) × (true concentration of standard sample)

Using the data in Table 4.4 and the above correction formula we obtain the corrected determinations in Table 5.6.

Table 5.6   Determinations corrected for relative bias

	Batches							
	A	*B*	*C*	*D*	*E*	*F*	*Mean*	*SD*
Determinations on unknown samples	20.6	20.5	20.7	20.6	20.8	21.0	20.70	0.179
Determinations on standard samples	10.3	10.7	10.6	11.0	10.3	10.4	10.55	0.274
Corrected determinations	20.0	19.2	19.5	18.7	20.2	20.2	19.63	0.609

When we compare the corrected determinations above with those in Table 4.4, how are we to decide which correction formula has been most effective? There is no way of discovering, from an inspection of the two tables, which set of corrected determinations is closer to the true concentrations of the unknown samples. We note that the standard deviation in the above table (0.609) is greater than that in the other table (0.394), but neither of these standard deviations is a measure of precision for Brown was not making repeat determinations on the same sample.

The only support that we have for using the new correction formula comes from the regression analysis which indicated that Brown had a relative bias but no significant fixed bias. If this finding is correct then the systematic errors in Brown's determinations should have been greatly reduced by the correction. The price we must pay for this gain is an amplification of the random errors.

5.6   Correlation and percentage fit

In this chapter we have introduced a very useful statistical technique – regression analysis. The full power of this technique has not been revealed because we applied it to a very simple problem. For a thorough treatment of the use of multiple regression analysis in research and development problems the reader is referred to Caulcutt (1982) or to Davies and Goldsmith (1972). In such texts the reader will meet an associated technique known as 'correlation', and this chapter would not be complete without a brief description of its use.

For any set of data to which regression analysis can be applied we could also calculate a correlation coefficient using the formula:

$$\text{Correlation coefficient} = \frac{Sxy}{\sqrt{(Sxx\ Syy)}}$$

For the data in Table 5.1 we have already used Sxy and Sxx to calculate the slope of the regression line. Syy is equal to $(n-1)(\text{SD of } y)^2$ which, for the data in Table 5.1, is equal to 279.940.

$$\text{Correlation coefficient} = \frac{264.5}{\sqrt{[(250.0)\,(279.94)]}}$$

$$= 0.999\,823\,16$$

The correlation coefficient tells us how well the regression line fits the points. A correlation of $+1.0$ or a correlation of -1.0 indicates a perfect fit, as we see in Figs 5.5(a) and 5.5(b). A correlation of 0.0, halfway between these two extreme values, may result from the random scatter of Fig. 5.5(c) or from the curved relationship in Fig. 5.5(d). Correlation is a measure of *linear association* between two variables.

When using regression analysis it is also useful to quote the *percentage fit* of the regression equation. This is closely related to the correlation coefficient:

$$\text{Percentage fit} = 100\,(\text{correlation coefficient})^2$$

For Figs 5.5(a) to 5.5(d) the percentage fits are 100%, 100%, 0% and 0%. For the data in Table 5.1 the percentage fit is $100(0.999\,823\,16)^2$ which is equal to 99.9646%. Thus the sample to sample variation in true concentration (x) accounts for 99.9646% of the variation in measured concentration. This is a very impressive percentage, but the statement is simply telling us that the errors introduced by the test method are very small compared with the large

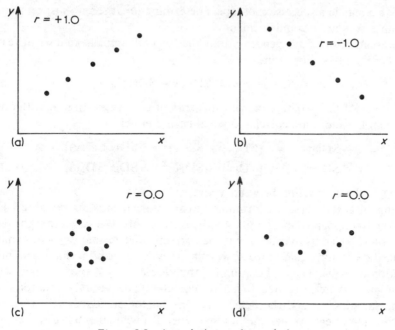

Figure 5.5   Association and correlation

variation between the samples. The analyst is searching for a needle in a haystack, but the needle must be found and then it must be measured.

Earlier in this chapter we tested for the presence of a variable bias by calculating a confidence interval for the true slope and then checking whether or not this interval included 1.0. A true slope of 1.0 was very meaningful in that particular situation. In many applications of regression analysis we would wish to answer the question 'are the two variables related?' which is equivalent to asking 'is the true slope equal to zero?'

To answer this question we could calculate a confidence interval for the true slope then check whether or not this interval included zero. A more convenient test, which will always give the same result, consists of calculating the correlation coefficient and comparing its value with the critical values in Table D. Many pocket calculators will fit a regression equation and calculate a correlation coefficient but few will give confidence intervals.

5.7   Correlated errors

We have assumed in earlier chapters that the random errors in successive determinations were *independent* of each other. Having described random errors as being 'unpredictable', it is necessary that they should not be dependent on each other. If, for example, we find long runs of positive errors and long runs of negative errors in a succession of repeat determinations then we suspect that the random errors are not independent.

The assumption of independence underlies several formulae that we have used in earlier chapters. For example:

$$SD(x+y) = \sqrt{[SD(x)^2 + SD(y)^2]} \tag{5.1}$$

is only valid if the variation in x is independent of the variation in y. If, on the other hand x and y are correlated we must use either:

$$SD(x-y) = \sqrt{[SD(x)^2 + SD(y)^2 - 2rSD(x)SD(y)]} \tag{5.2}$$

or $\qquad SD(x+y) = \sqrt{[SD(x)^2 + SD(y)^2 + 2rSD(x)SD(y)]} \tag{5.3}$

where r is the correlation between x and y.

Using the first of these two formulae we can return to the data in Table 4.4 and explain what appeared in Chapter 4 to be an anomaly. We noted that the use of Equation (5.1) led us to believe that the corrected determinations would have a standard deviation equal to 0.327 whereas they were found to have a standard deviation of 0.394. The explanation for this discrepancy is that the two sets of determinations are correlated. In fact the correlation between the determinations on the samples (x) and the determinations on the standards (y) is equal to -0.4899. (The negative sign implies that when Brown makes a positive error on the sample he tends to make a negative error on the standard within the same batch. One wonders why this should be so.) Using Equation (5.2) we get:

SD(corrected determinations)

$$= \sqrt{[SD(samples)^2 + SD(standards)^2 - 2r\,SD(samples)SD(standards)]}$$
$$= \sqrt{[(0.179)^2 + (0.274)^2 - 2(-0.4899)(0.179)(0.274)]}$$
$$= 0.394$$

This result is in agreement with the standard deviation of the corrected determinations in Table 4.4.

5.8 Summary

In this chapter we concentrated on relative bias in which systematic errors are proportional to the concentration being measured. We used simple regression analysis in an attempt to separate the fixed errors and the proportional errors in a set of determinations made on samples with different concentrations. The task was made more difficult by the presence of random errors of course.

In addition to the 'best straight line through the centroid' we also fitted a 'best straight line through the origin'. Both methods of fitting a line will prove useful when we discuss calibration lines in Chapter 7. Before we set foot in that important field of statistical application, however, we must give further consideration to the estimation of precision and to ways of exploring the possibility that the precision of a test method might be related to the concentration being measured.

Problems

(1) To estimate the fixed and relative bias of a method six samples of different known concentrations were prepared and measured using an analytical method. Results are given in the table below.

(a) Complete the following table and hence calculate the slope and intercept.

True concentration (x)	Measured concentration (y)	(x−x̄)	(y−ȳ)	(x−x̄)(y−ȳ)	(x−x̄)²
10	11.0	−25	−25.7	642.5	625
20	21.3	−15	−15.4	231.0	225
30	31.5	−5	−5.2	26.0	25
40	41.9	5	5.2	26.0	25
50	51.9				
60	62.6				
x̄ = 35	ȳ = 36.7				

(b) Complete the following table:

True concentration	Estimated error due to fixed bias	Estimated error due to relative bias	Estimated total error due to bias	Predicted determination	Measured concentration (y)	Residual (estimated random error)	Squared residual
10	0.68	0.29	0.97	10.97	11.0	0.03	0.0009
20	0.68	0.58	1.26	21.26	21.3	0.04	0.0016
30	0.68	0.87	1.55	31.55	31.5	−0.05	0.0025
40	0.68	1.16	1.84	41.84	41.9	0.06	0.0036
50					51.9		
60					62.6		

(c) Calculate the residual standard deviation.
(d) Calculate 95% confidence intervals for the intercept and slope.
(e) Is there a significant fixed and/or relative bias present?

(2) A regression line through the origin to the data in the first problem gives the following statistics:

$$b = \frac{\sum xy}{\sum x^2} = \frac{9508}{9100} = 1.045$$

$$RSD = 0.354$$

95% confidence interval for the true slope:

$$1.045 \pm 0.0095$$

Comment on the higher RSD and the smaller confidence interval for the slope.

(3) Based on the results of the first problem it is decided that a standard and possibly a blank should be included in each batch and corrected determinations calculated from one of the following formulae:

(a) $\left(\begin{array}{c}\text{Corrected}\\\text{determination}\end{array}\right) =$

$$\left(\begin{array}{c}\text{determination}\\\text{of}\\\text{unknown}\\\text{sample}\end{array}\right) - \left(\begin{array}{c}\text{determination}\\\text{of}\\\text{standard}\end{array}\right) + \left(\begin{array}{c}\text{true}\\\text{concentration}\\\text{of}\\\text{standard}\end{array}\right)$$

(b) $\left(\begin{array}{c}\text{Corrected}\\\text{determination}\end{array}\right) =$

$$\left(\begin{array}{c}\text{determination}\\\text{of}\\\text{unknown}\\\text{sample}\end{array}\right) \times \left(\begin{array}{c}\text{true}\\\text{concentration}\\\text{of}\\\text{standard}\end{array}\right) \Bigg/ \left(\begin{array}{c}\text{determination}\\\text{of}\\\text{standard}\end{array}\right)$$

(c) $\left(\begin{array}{c}\text{Corrected}\\\text{determination}\end{array}\right) =$

$$\frac{\left[\left(\begin{array}{c}\text{determination}\\\text{of}\\\text{unknown}\\\text{sample}\end{array}\right) - \left(\begin{array}{c}\text{determination}\\\text{of}\\\text{blank}\end{array}\right)\right] \times \left(\begin{array}{c}\text{true}\\\text{concentration}\\\text{of}\\\text{standard}\end{array}\right)}{\left(\begin{array}{c}\text{determination of}\\\text{standard}\end{array}\right) - \left(\begin{array}{c}\text{determination of}\\\text{blank}\end{array}\right)}$$

Consider determinations made on unknown samples at $\frac{1}{2}$, 1 and $1\frac{1}{2}$ times the concentration of the standard and hence decide which method of correction is preferable.

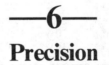

Precision

6.1 Introduction

In later chapters we will discuss 'precision experiments'. These are carried out with the purpose of estimating the repeatability and reproducibility of a standard test method. In that discussion we will explore the recommendations of two British Standards which offer guidance on the conduct of precision experiments and we will analyse the results of such an experiment.

Before we can undertake such a complex task, however, we must have a clear perception of how the random variation associated with a test can be quantified, and how this random variation may depend upon the use of different operators in different situations. As a first step in this direction we will in this chapter examine methods of estimating and comparing standard deviations. As the standard deviation of a set of determinations can be greatly inflated by the occurrence of a gross error in one determination, we will first explore a 'test for outliers' which can be used to detect such errors. We will then go on to calculate confidence intervals for standard deviations and to examine ways to compare one laboratory/operator with another or to decide whether the worst laboratory of a group is capable of matching the standard of precision set by the others.

6.1 Estimating precision

The reader will recall that, in Chapter 3, we compared the performance of four analysts. Each was asked to make six determinations of the ammonia content of a consignment of ammonia solution. The results are reproduced in Table 6.1.

We have already explored several features of this data. We concluded in Chapter 4 that Brown was biased. Further investigation in Chapter 5 revealed that Brown's bias was related to the concentration being measured, and his systematic error was estimated to be approximately 6%. Let us assume that suitable training would result in this bias being eliminated or reduced to an insignificant level. Brown would then be comparable with Smith, whom we have already declared to be acceptable. Jones and Lee appear not to be biased but both have larger standard deviations than Brown or Smith. In Miss Lee's case we have

Table 6.1 Repeat determinations by four analysts

Analyst	Determinations of ammonia content						Mean	SD
Smith	20.2	19.9	20.1	20.4	20.2	20.4	20.20	0.190
Jones	19.9	20.2	19.5	20.4	20.6	19.4	20.00	0.486
Brown	20.6	20.5	20.7	20.6	20.8	21.0	20.70	0.179
Lee	20.1	19.9	20.2	19.9	21.1	20.0	20.20	0.456

already noted that one of her determinations (21.1) does not appear to fit in with the pattern exhibited by the other five. This was particularly noticeable when we examined a blob chart of Miss Lee's data. If we were to eliminate this outlying observation the mean and standard deviation of the remaining determinations would be 20.02 and 0.130 respectively. Thus the rejection of this 'rogue' result would bring Miss Lee very much into line with Smith and Brown.

On what grounds could the rejection of this one determination be justified? Surely we cannot reject an apparent outlier simply because this course of action helps us to confirm a preconceived opinion, that Miss Lee is as good as any man in the laboratory. A sound basis for the rejection of an outlying determination would lie in the discovery that abnormal conditions existed at the time that this particular analysis was being carried out. Perhaps, on reflection, we realize that some temporary characteristic of the analyst or the environment may have resulted in a gross error contributing to this determination.

If a 'post-mortem' investigation failed to reveal any abnormality, we could not be certain, of course, that conditions were perfectly standard at the time the analysis took place. Perhaps the *mere possibility* of an undetected abnormality would justify the use of a statistical technique, as an objective basis for deciding whether or not the 21.1 is an outlier. Many such techniques exist under the generic heading 'tests for outliers', and one of the most commonly used is Dixon's test. We will make use of Dixon's test to help us decide, on purely statistical grounds, whether it is reasonable to conclude that the outlying result does not belong within the scatter of determinations we would expect to find when Miss Lee is performing 'normally'. The test we carry out will be based on the assumption that Miss Lee's determinations have a normal distribution. We will attempt to answer the question 'Is it likely that the sixth determination (21.1) came from the *same* distribution as the other five?'

Null hypothesis – All six determinations came from the same normal distribution.
Alternative hypothesis – Either the highest or lowest determination did not come from the same normal distribution as the other five.
Test statistic = the greater of *A* or *B*:

$$A = \frac{x_2 - x_1}{x_6 - x_1}; \qquad B = \frac{x_6 - x_5}{x_6 - x_1}$$

$(x_1 = 19.9;\quad x_2 = 19.9;\quad x_3 = 20.0;\quad x_4 = 20.1;\quad x_5 = 20.2;\quad x_6 = 21.1)$

$$A = \frac{19.9 - 19.9}{21.1 - 19.9};\qquad B = \frac{21.1 - 20.2}{21.1 - 19.9}$$

$$A = 0.0;\qquad\qquad B = 0.75$$

$$\text{Test statistic} = 0.75$$

Critical values – from Table G for a sample size of 6 are:
　　　　0.628 at the 5% significance level
　　　　0.740 at the 1% significance level
Decision – We reject the null hypothesis at the 1% level of significance.
Conclusion – We conclude that the largest determination (21.1) does not belong
　　　　to the same normal distribution as the other five determinations.

We have proved, beyond reasonable doubt, that the outlying determination is
not in keeping with Miss Lee's normal performance as indicated by the other five
determinations. Several courses of action are now open to us, one of which is to
reject the outlier. For a full discussion of the testing and treatment of outliers the
reader is referred to Barnett and Lewis (1978).

Let us reject the rogue value and use only the other five determinations. The
performance of the four analysts is then summarized by the means and standard
deviations in Table 6.2.

Table 6.2　Precision of the four analysts

Analyst	No. of determinations	Mean	SD
Smith	6	20.20	0.190
Jones	6	20.00	0.486
Brown	6	20.70	0.179
Lee	5	20.02	0.130

What conclusions can we draw concerning the precision of the four analysts? Is
Miss Lee the most precise? Are Smith, Brown and Lee equally precise, with Jones
being distinctly inferior? In answering such questions we would need to bear in
mind that each of the four standard deviations is based on only a *sample* of
determinations whereas the above questions implicitly refer to the *populations*
from which these samples were taken. To draw a conclusion about a population
using data from a sample we can, of course, use a significance test. Before doing so
we will calculate a confidence interval for each of the population standard
deviations.

A two-sided confidence interval for a population standard deviation is given by:

$$\text{Lower limit} = L_1 s$$
$$\text{Upper limit} = L_2 s$$

where L_1 and L_2 are obtained from Table F and s is the sample standard deviation

Smith's standard deviation (0.190) was calculated from six determinations and has, therefore, 5 degrees of freedom. With 5 degrees of freedom the values from Table F are $L_1 = 0.62$ and $L_2 = 2.45$, for a 95% confidence interval. We can therefore be 95% confident that Smith's long-term standard deviation would lie between 0.118 (i.e. 0.62×0.190) and 0.466 (i.e. 2.45×0.190). Carrying out similar calculations for the other three analysts we obtain the confidence intervals in Table 6.3.

Table 6.3 95% confidence intervals for population standard deviation

Analyst	Confidence interval		
Smith	0.118	to	0.466
Jones	0.301	to	1.191
Brown	0.111	to	0.439
Lee	0.078	to	0.373

The confidence intervals in Table 6.3 are presented pictorially in Fig. 6.1. We note that each of the intervals is asymmetrical, unlike the confidence intervals for population means that we calculated in Chapter 4. We also note that all four intervals overlap. It would, however, be unwise to jump to the conclusion that all four population standard deviations are equal. Checking for overlap of two or more 95% confidence intervals is *not* equivalent to carrying out an appropriate significance test using a 5% significance level.

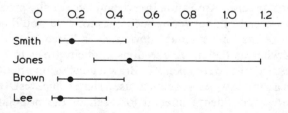

Figure 6.1 95% confidence intervals for population SD

6.3 The number of determinations needed to estimate precision

In Chapter 4 we asked the question 'How many determinations are needed to estimate a fixed *bias* to within $\pm c$?' In answer to this question we offered a formula:

$$n = \left(\frac{ts}{c}\right)^2$$

which could be used to calculate the necessary sample size. It might seem reasonable to ask a similar question concerning the estimation of precision. Unfortunately there are fundamental differences between a confidence interval for a population mean and a confidence interval for a population standard deviation:

(a) The former is symmetrical whereas the latter is not.
(b) The former is calculated by *adding* and *subtracting* a 'half width' whereas the latter is calculated by multiplying.

To speak of 'estimating a population standard deviation to within $\pm c$' would therefore be meaningless. We can, however, obtain sample size formulae if we confine our attention to *one-sided* confidence intervals. Such intervals are calculated using L_1 and L_2 values from Table F, which is the table that we used when calculating two-sided confidence intervals.

A lower confidence limit for a population standard deviation is equal to $L_1 s$.

An upper confidence limit for a population standard deviation is equal to $L_2 s$.

Where L_1 or L_2 is obtained from Table F and s is the sample standard deviation.

If we wished, for example, to calculate a 95% upper confidence limit for the long term standard deviation of Mr Smith, we would use an L_2 value of 2.09 from Table F. This is taken from the one-sided upper 95% column using 5 degrees of freedom. Smith's sample standard deviation was 0.190, giving an upper limit of 2.09×0.190 which is 0.397. We can therefore be 95% confident that, if Smith had made many many determinations on the standard sample, his standard deviation would have been less than 0.397.

We could also calculate a 95% lower confidence limit. This would be equal to 0.67×0.190 which is 0.127. We can therefore be 95% confident that Smith's long-term standard deviation would be greater than 0.127. Note that these two limits, when taken together, do *not* constitute a two-sided 95% confidence interval but they *do* give a two-sided 90% confidence interval. Thus we can be 90% confident that Smith's population standard deviation lies between 0.127 and 0.397. This interval is of course narrower than the 95% interval quoted in Table 6.3. The reader will realize that the decision to calculate a one-sided confidence interval, rather than the more usual two-sided interval, would depend upon the situation and the question being asked. Ideally such a decision would be made *before* the data was gathered.

By using Table F in reverse we can make predictions about sample sizes. For example, if we examine the 95% one-tail column of Table F we find that the largest entry which is less than 2.00 is the 1.92 corresponding to 6 degrees of freedom. We can conclude therefore that we need at least 6 degrees of freedom if we are estimating a population standard deviation and we wish to obtain an upper 95% confidence limit that is less than twice the estimate. If this estimate were the standard deviation of n repeat determinations then n would need to be at least 7. Alternatively we would need 18 degrees of freedom if we required a 95% upper limit that was approximately 40% greater than the estimate.

If the reader expected a statement of the form 'n observations are needed in order to estimate the precision of an instrument to within ± 0.001' he may be disappointed. We will however have more to say about sample sizes later in this chapter.

6.4   Is Jones less precise than the other analysts?

When comparing the precision of several analysts or several test methods or several laboratories, there are various comparisons that we might make. Often we are particularly interested in comparing the *worst* analyst, or method, or laboratory with the others. The purpose of such a comparison might be to single out the worst analyst for special attention or it might be a necessary prerequisite to further statistical analysis, as we shall see later.

Most statistical tests concerned with variability are based on *variances* rather than standard deviations, and we will now make use of Cochran's test which compares the *largest* of a set of variances with the other variances in the set. Clearly this test can be used to check the significance of the most variable analyst or the most variable laboratory or the most variable test method. Cochran's test is one of several significance tests which are known as *homogeneity of variance* tests and which share the null hypothesis that 'several methods (or analysts or laboratories) are equally variable'.

Null hypothesis – The four analysts are equally precise.
Alternative hypothesis – The most variable analyst is less precise than the other three analysts.

$$\text{Test statistic} = \frac{\text{largest variance}}{\text{sum of all four variances}}$$

$$= \frac{(0.486)^2}{(0.190)^2 + (0.486)^2 + (0.179)^2 + (0.130)^2}$$

$$= 0.735$$

Critical values – from Table H, for four 'laboratories' with six determinations in each, are:

0.590 at the 5% significance level
0.676 at the 1% significance level

Decision – We reject the null hypothesis at the 1% level of significance.
Conclusion – We conclude that Jones is more variable (i.e. less precise) than the other three analysts.

Cochran's test is telling us that the excessive variability in Jones's six determinations cannot reasonably be written off as an unsual occurrence. There is very little chance that an analyst as precise as Smith, Brown or Lee would produce six repeat determinations with a standard deviation as large as that exhibited by Jones (0.486).

Cochran's test is designed for use in situations where several laboratories (or analysts or methods) have produced equal numbers of repeat determinations. When referring to Table H we assumed that all four variances were based on six determinations, though this was not exactly true since one of Miss Lee's results had been discarded. The error introduced by this deviation would be small and would not have affected our decision. To overcome this difficulty, we could have used a method for replacing missing values. These methods can be very useful when the data from a carefully balanced experiment is spoiled by a small number of missing or rejected determinations.

Having concluded that Jones cannot match the precision of the other three analysts we could now identify the next largest standard deviation in Table 6.2 (0.190 for Smith) and apply Cochran's test again. If we did so we would not be able to reject the null hypothesis and would not therefore conclude that Smith was inferior to Brown and Lee.

6.5 Combining standard deviations

Having failed to prove that the three analysts differ in precision it is reasonable to combine the standard deviations of Smith, Brown and Lee. This would give us an estimate of the precision of the test method when being used by an acceptable analyst. The usefulness of this single estimate would depend of course on how representative Smith, Brown and Lee are of the larger group of analysts who might use this test method. If these three are very superior whilst the rejected Jones is closer to the average analyst, then our combined estimate may be a target that few analysts could reach.

The reader may recall the assertion made in Chapter 3 that 'we must *never* add standard deviations, but we *can* add variances'. In calculating a combined estimate of precision, therefore, we will take the square root of the mean of the variances. A further complication is introduced because Miss Lee's standard deviation is based on fewer determinations than that of Smith or Brown. For this reason we will use a weighted average rather than a straight average, with each variance being weighted by its degrees of freedom.

$$\text{Combined standard deviation} = \sqrt{\left\{\frac{\sum[(d.f.)(SD)^2]}{\sum(d.f.)}\right\}}$$

$$= \sqrt{\left[\frac{5(0.190)^2 + 5(0.179)^2 + 4(0.130)^2}{5+5+4}\right]}$$

$$= 0.171$$

This combined estimate (0.171) must, of course, lie between the smallest standard deviation (0.130) and the largest (0.190), no matter how different are the degrees of freedom of the standard deviations. Combined estimates of variability or precision can be very important in analytical chemistry. With the widespread practice of carrying out routine analyses in batches on automatic equipment, the inclusion of 'standards' and/or 'blanks' in every batch will allow us to calculate a separate precision estimate for each batch. Unfortunately each estimate is of little use on its own for it is based on very few degrees of freedom but a combined estimate based on several batches is extremely useful. It can be used for checking the bias or calibration of the method during each successive batch.

6.6   Is the new operator acceptable?

Our combined standard deviation (0.171) based on 14 degrees of freedom could be used as a standard against which to compare the precision of a new operator. We could ask the new operator to make several repeat determinations of the concentration of a standard solution. If his standard deviation were significantly greater than 0.171 then we could declare him to be unacceptable. If, on the other hand, he appeared to be equal to or better than the 'standard analyst' we could welcome him into the fold.

To help us reach a decision on the precision of the new operator we could carry out a *one-sided* significance test. In Chapter 4 we carried out a *two-sided* t-test because we were interested in detecting a bias in *either* direction. A positive bias or a negative bias would have been equally bad and it was important to detect either, if it existed. In the present situation we are looking for a difference in *one* direction only; we are asking 'Is the new operator *worse* than the standard?'

Suppose the new operator were asked to make ten repeat determinations on a standard solution known to contain 20% ammonia and his determinations were:

20.5	19.9	20.1	20.3	19.9
19.7	20.2	19.6	19.7	20.1

The mean of these ten determinations is 20.00, so we have no evidence that the new operator is biased. The standard deviation of the determinations is 0.2906, which is certainly greater than the 0.171 that we are using as a standard. To see if the new operator's standard deviation is *significantly* greater than the standard we will carry out a one-sided F-test as follows:

Null hypothesis – The precision of the new operator is equal to the standard (i.e. $\sigma = 0.171$).

Alternative hypothesis – The precision of the new operator is inferior to the standard (i.e. the operator is more variable, $\sigma > 0.171$).

Test statistic = larger variance/smaller variance

$$= (0.2906)^2/(0.171)^2$$

$$= 2.89$$

Critical values – from the one-sided F-table (Table C) with 9 and 14 degrees of freedom are:

2.65 at the 5% significance level
4.03 at the 1% significance level.

Decision – We reject the null hypothesis at the 5% significance level.

Conclusion – We conclude that the new operator is more variable than the standard against which he was compared.

We are led to the conclusion that the new operator does *not* match up to the standard set by Smith, Brown and Lee. The scatter of his ten determinations is excessive. A graphical inspection and/or a Dixon's test would fail to reveal any outliers, so the variability exhibited by the new operator cannot be accounted for by one rogue determination. It should be pointed out that the F-test is based on the assumption that both populations have a normal distribution. Furthermore, the F-test is rather more sensitive to the violation of this assumption than is the t-test. A visual inspection of the data, however, gives no clear indication that either the determinations of the new operator or those of the three analysts came from a distribution that wasn't normal.

6.7 The number of determinations needed to compare two analytical methods

In the previous section we used the F-test to compare the variability, or the precision, of two analysts. We could obviously follow a similar procedure in order to compare the precision of two analytical methods. Perhaps we wish to decide which of two available methods we should use for a particular purpose. A number of repeat determinations by each method would furnish data for a two-sided F-test. Perhaps, on the other hand, we wish to compare a new analytical method with an existing method and we are willing to adopt the new method only if it is demonstrated to be more precise than the old. In this case we would carry out a one-sided F-test on repeat determinations made by the two methods.

In both situations we would need to decide how many repeat determinations should be made by each method. Obviously the larger the number of determinations the more confident we will be that we have selected the better method. No matter how many determinations are made, however, there will always be a chance that we will select the worse of the two methods. Naturally we wish to reduce this chance by making a sufficiently large number of determinations but only by testing indefinitely can we reduce the chance to zero. A further point to note is that we are more likely to detect the better method if there is a substantial difference between the two. When choosing sample sizes, therefore we need to take into account the confidence level and the percentage difference that we desire to detect.

> To be confident of choosing the more precise of two methods (or analysts), when the other is $x\%$ more variable, we choose degrees of freedom in Table C which give an F-value less than $(1 + x/100)$.

Suppose, for example, that we are comparing two existing analytical methods and we wish to be 95% confident of selecting the better of the two methods if the standard deviation of the other is 200% greater (i.e. three times as variable). Letting $x = 200$ we find that $(1 + x/100)$ is equal to 3.00. We must now find an F-value less than 3.00 in the 5% two-sided section of Table C. Clearly there are several possibilities. The two estimates need not have the same degrees of freedom, but the total number of tests will be minimized if they do. If we decide therefore to use equal sample sizes we see that 15 degrees of freedom for each gives an F-value of 2.86, which is less than 3.00. Sixteen determinations by each method will therefore give us at least 95% chance of detecting the better method, if the other method is 200% more variable. When the determinations have been made we will use a two-sided F-test with the null hypothesis that the two methods are equally variable. If there really is a 200% difference we run a 5% risk of failing to reject this hypothesis. In other words we run a 5% risk of failing to detect such a large difference if it actually exists. Unfortunately this is not the *only* risk. By using the recommended sample size ($n = 16$) we will also run an equal risk (i.e. 5% in this case) of concluding that 'one method is 200% better than the other' if they are actually equally precise. In the language of the statistician, the risks of type I error and of type II errors are both equal to 5% if we make sixteen determinations by each method.

If we wished to detect a smaller difference, say 50%, then larger sample sizes would be needed. Furthermore, if we wished to be more confident (say 99%) of detecting a difference large sample sizes would again be needed.

6.8 Is the precision of the method related to the concentration?

In Chapter 5 we discussed the possibility that the bias of an analyst or a method might be either fixed or relative or a combination of the two. We used simple regression analysis to split the total bias of a particular analyst (Brown) into its fixed and variable components. We eventually concluded that Brown had a relative bias of approximately 6%.

Let us now turn our attention to the possibility that the *precision* of an analyst (or of a test method) might be related to the concentration of the determinand. Suppose that one of our unbiased analysts, Smith, had been asked to make six repeat determinations at several different concentrations of ammonia. The results of this experiment are listed in Table 6.4.

Table 6.4 Repeat determinations at five concentrations

	Approximate true concentration				
	10%	15%	20%	25%	30%
Determinations	9.9	15.2	20.2	24.7	30.6
	9.9	15.1	19.9	25.0	29.8
	10.0	15.1	20.1	25.1	30.2
	9.8	15.1	20.4	25.0	30.2
	9.7	14.9	20.2	24.9	30.5
	10.1	15.2	20.4	24.7	29.9
Mean	9.90	15.10	20.20	24.90	30.20
Standard deviation	0.141	0.110	0.190	0.167	0.316
Coefficient of variation	1.42%	0.74%	0.94%	0.67%	1.05%

We see in Table 6.4 that Smith's repeat determinations tend to be more variable at higher concentrations. This relationship is also apparent in Fig. 6.2 where the standard deviation of each set of determinations is plotted against the mean.

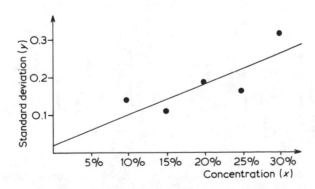

Figure 6.2 Is precision related to concentration?

Obviously a regression line fitted to the points in Fig. 6.2 would have a positive slope. In fact the equation of the best straight line is:

$$y = 0.022 + 0.008\,11x$$

where y = standard deviation and x = mean.

The intercept of the regression line (0.022) suggests that repeat determinations at zero concentration can be expected to have a standard deviation of 0.022. The slope of the regression line (0.008 11) suggests that the standard deviation can be expected to increase by 0.008 11 for every 1% increase in concentration.

We can, of course, calculate confidence intervals for the true slope and the true intercept. Following the procedure outlined in Chapter 5 we obtain a residual standard deviation of 0.052 771 and a 95% confidence interval for the true intercept which extends from -0.20 to $+0.25$. On statistical grounds, therefore, it is quite possible that the true intercept is equal to zero. Analytical consideration might of course rule out this possibility entirely. It is quite unlikely that random errors introduced by the test method will disappear completely when we attempt to measure a concentration that has a true value of zero. Because the lowest concentration in our experiment was 10%, we are unable to make an accurate estimate of the precision at zero concentration. Let us turn our attention therefore to the relationship between precision and concentration.

A 95% confidence interval for the true slope of the line in Fig. 6.2 is 0.0081 ± 0.0105, which extends from -0.0024 to $+0.0186$. Because this interval includes zero we are unable to reject the null hypothesis that the true slope is equal to zero. We are unable to conclude, therefore, that the precision of the test method *is* related to the concentration being measured. The relationship that we see in Fig. 6.2 could simply be due to chance and appears not therefore to be a manifestation of a real characteristic of the test method.

We can now reach a decision concerning the 'error structure' of the test method. It appears that, when being used by Dr Smith, the test method gives rise to determinations resembling those in Fig. 6.3(a) rather than those in Fig. 6.3(b). Note that *neither* of these diagrams is based on the data in Table 6.4. In fact Fig. 6.3 is drawn 'larger than life' to illustrate the difference between the two error

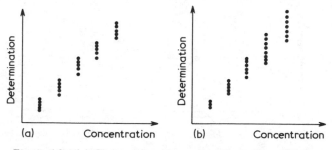

Figure 6.3   (a) Constant precision, (b) decreasing precision

structures. Figure 6.3(a) represents a test method with constant precision, whilst Fig. 6.3(b) represents a test method for which variability increases with concentration.

Obviously many test methods conform to Fig. 6.3(b) rather than Fig. 6.3(a). Furthermore, it is very important that we should know which of the two diagrams best describes the error structure of a test method *before* we attempt to fit a calibration line. Unfortunately the method of least squares is based on the assumption that the 'error standard deviation' is constant. We can therefore use the method with data resembling that in Fig. 6.3(a) but we must use what is known as 'weighted least squares' or 'weighted regression' if we wish to fit a straight line to data resembling that in Fig. 6.3(b). This point will be discussed further when we explore calibration problems in Chapter 8.

6.9 Summary

In this chapter we have discussed the *precision* of test methods. Using repeat determinations on a standard solution we calculated confidence intervals for the 'long term' standard deviations of four analysts using the same test method. Such estimates can be considerably influenced by a gross error in one or more determinations, so we first used Dixon's Test to check for outliers.

Later we used Cochran's Test to see if any analyst was significantly less precise than the others. Cochran's Test could be regarded as an outlier test for analysts (or laboratories) and it will be used as such in a later chapter. We also made use of the F-test to compare two operators or to compare one operator with a required standard of precision. Unfortunately, rather large samples are needed in order to estimate or check precision as we found when we calculated the size of sample needed to compare two test methods. We saw that sixteen repeat determinations were needed with each method if we were to have a 95% chance of detecting a 200% difference. This requirement for large samples is perhaps better seen as a requirement for a large number of degrees of freedom, which can be obtained by combining standard deviations obtained from several small samples.

Finally we considered the possibility that the precision of a test method might not be constant but might be related to the level of determinand. This possibility is important in calibration which will be considered in the next two chapters.

Problems

(1) In a previous problem Green obtained eight determinations on a sample containing a known concentration of chloride in water. His determinations were:

    49.4   49.8   50.8   49.3   51.3   50.0   50.8   51.8

(a) Calculate a 95% confidence interval for Green's long term standard deviation.

(b) In the above calculation it is assumed that all eight determinations came from the same normal distribution. Check this assumption.

(2) To characterize the above analytical method seven experienced operators have carried out a number of replicate determinations on unknown samples. The results are:

Operator	Determinations				SD
A	49.1	48.5	49.5	49.0	0.411
B	57.4	59.2	57.8	58.5	0.793
C	44.2	43.6	43.7	45.1	0.685
D	61.0	60.0	59.8	61.2	0.702
E	53.1	53.6	54.1	53.9	0.434
F	41.2	40.1	41.0		0.585
G	63.1	61.6	62.4	61.4	0.780

(a) Calculate a combined SD from the determinations. We shall refer to this value as the SD of the experienced operators.
(b) Test whether Green is less precise than the experienced operators.
(c) How less precise would Green need to be before we could be confident of detecting his imprecision using an F-test?

—7—
Calibration

7.1 Introduction

In previous chapters many sets of determinations have been examined. In every case it was assumed that a standardized test method had been used and where the test procedure required calibration it was further assumed that this had been carried out. No mention of calibration was made in earlier chapters because it was intended that the statistical problems associated with calibration would be dealt with separately.

By removing calibration from the main stream of the book and by deferring its discussion to later chapters we may have given the impression that errors occurring during calibration are:

(a) unimportant;
(b) unrelated to errors which arise during the direct use of the test method.

Nothing could be further from the truth, for the factors which introduce random and systematic errors during operational use of the test method, will probably be present at the calibration stage. It is possible, of course, that the calibration standards will have a matrix that differs somewhat from that of the samples, in which case the bias and precision during calibration may differ from the bias and precision during operation. Whether or not this is so, errors during calibration cannot be ignored.

In this chapter we will explore several aspects of calibration but we will confine our attention to situations in which the precision of the test method *is not related to the concentration of the determinand*. We will compare the fitting of a calibration line through the centroid ($y = a + bx$) with a calibration line through the origin ($y = bx$). In both cases we will calculate a confidence interval for the true concentration of an unknown sample and we will consider strategies for reducing the width of this interval. Finally we will explore a method of correcting for bias at the calibration stage.

7.2   Fitting a calibration line

The four analysts (Smith, Jones, Brown and Lee) referred to in earlier chapters were employed by Indichem Ltd. This company uses large quantities of ammonia solution which is purchased from several suppliers. A new and inexpensive source of supply has arisen, but the Chief Chemist of Indichem suspects that ammonia solution from this source might be contaminated with cuprammonium. He therefore devises a spectrophotometric method of determining the concentration of cuprammonium ion in ammonia solution. This involves measuring the absorbance of a sample at a wavelength of 600 nm.

Undoubtedly there is a relationship between the absorbance reading of the spectrophotometer and the concentration of cuprammonium ion in the sample. This relationship may be influenced by the presence of other compounds in the sample, of course, but the Chief Chemist intends to defer the investigation of these interference effects until later. His immediate requirement is to calibrate the test method by quantifying the relationship between absorbance (y) and concentration (x), using a range of concentration which is likely to be encountered when monitoring deliveries. In order to obtain data which can be used to establish a calibration curve the Chief Chemist prepares five samples of known concentration and records the absorbance of each.

Table 7.1   Absorbance of five 'known' samples

Concentration (M)	x	0.002	0.003	0.005	0.008	0.012
Absorbance	y	0.12	0.14	0.27	0.40	0.52

The data in Table 7.1 has been plotted in Fig. 7.1. Clearly there is a very 'strong' relationship between the two variables; as there must be if the test method is to have any practical use. In order to quantify this relationship we will fit a least-squares regression equation of the form $y = a + bx$. The calculations will be identical to those carried out in Chapter 5 when we discussed relative bias. Using a pocket calculator we quickly obtain the following results:

$$\bar{x} = 0.006 \qquad SD(x) = 0.004\,062$$
$$\bar{y} = 0.290 \qquad SD(y) = 0.170\,88$$
$$Sxx = 0.000\,066 \qquad Syy = 0.1168 \qquad Sxy = 0.002\,75$$
$$\text{Slope } (b) = Sxy/Sxx$$
$$= 0.002\,75/0.000\,066$$
$$= 41.667$$
$$\text{Intercept } (a) = \bar{y} - b\bar{x}$$
$$= 0.290 - 41.667(0.006)$$
$$= 0.040$$

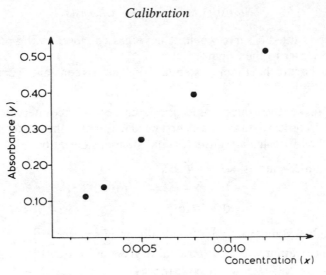

Figure 7.1 Data from calibration experiment

The equation of the least-squares 'best' straight line is $y = 0.040 + 41.667x$ and this line has been added to the data in Fig. 7.2. We see that the line passes through the centroid but does not pass through the origin. In fact the relationship of the line to the points in Fig. 7.2 raises several questions, which are somewhat interdependent:

(a) Should we have fitted an equation of the form $y = bx$ which would have forced the line through the origin?
(b) Why didn't the Chief Chemist include a sample which had zero concentration of cuprammonium (i.e. a blank)?

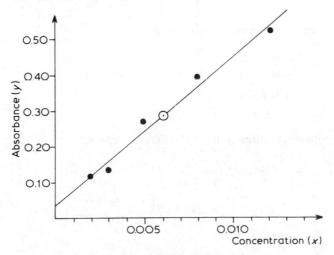

Figure 7.2 The calibration line

(c) If we had fitted a curve would it have passed closer to the points and/or passed closer to the origin?

(d) Can we be sure that the error standard deviation is constant over this range of concentration?

Before we answer these questions let us calculate confidence intervals as we did in Chapter 5. In order to do so we will first need to quantify the scatter of the points about the line by calculating the residual standard deviation.

$$\text{Residual sum-of-squares} = Syy - b^2 Sxx$$

$$= 0.1168 - (41.667)^2 (0.000\,066)$$

$$= 0.002\,216\,7$$

$$\frac{\text{Residual standard}}{\text{deviation}} = \sqrt{\left[\frac{\text{(residual sum of squares)}}{\text{(residual degrees of freedom)}}\right]}$$

$$= \sqrt{(0.002\,216\,7/3)}$$

$$= 0.027\,18$$

The residual standard deviation will be an estimate of the precision of the test method provided that we have fitted the right type of equation and the precision is not dependent on the concentration. We have fitted a straight line $y = 0.04 + 41.667x$ but the true relationship may be curved. The Beer–Lambert Law would suggest that the relationship between concentration and absorbance should be linear, certainly for low concentrations. If, however, the true relationship were *not* linear then the residual standard deviation would be increased and would give a false indication of the precision of the test method. We will return to this point in the next chapter, but for our present purposes, we will use the residual standard deviation as an estimate of the error standard deviation (ESD).

A 95% confidence interval for the true slope is:

$$b \pm t(\text{ESD})/\sqrt{(Sxx)}$$

$$= 41.667 \pm 3.18(0.027\,18)/\sqrt{(0.000\,066)}$$

$$= 41.667 \pm 10.639$$

$$\text{(i.e. } 31.03 \quad \text{to} \quad 52.31)$$

A 95% confidence interval for the true intercept is:

$$a \pm t(\text{ESD})\sqrt{\left(\frac{1}{n} + \frac{\bar{x}^2}{Sxx}\right)}$$

$$= 0.04 \pm 3.18(0.027\,18)\sqrt{\left[\frac{1}{5} + \frac{(0.006)^2}{0.000\,066}\right]}$$

$$= 0.04 \pm 0.075$$

$$\text{(i.e. } -0.035 \quad \text{to} \quad 0.115)$$

The reader may be surprised by the width of these confidence intervals. They are however a fair reflection of the confidence we can place in the estimates which are, we must remember, based on only *five* absorbance readings. The confidence interval for the true intercept extends from -0.035 to $+0.115$. As this interval includes zero, we are unable to reject the suggestion that the true relationship is of the form $y = bx$.

It would be quite reasonable therefore to fit a line which passed through the origin. Before doing so, let us consider how the calibration line will be used when we attempt to measure the concentration of a batch of ammonia solution.

7.3 Estimating the concentration of an unknown sample

Suppose, for example, a consignment of ammonia solution is sampled and the sample gives an absorbance reading of 0.20. To calculate a concentration (x) corresponding to this absorbance (y) we need to transpose our regression equation into the form:

$$x = (y - a)/b$$

i.e. $x = (y - 0.04)/41.667$

Substituting $y = 0.20$ gives $x = 0.003\,84$ and we conclude therefore that the concentration of cuprammonium ion in this batch is 0.003 84 M.

The reader will realize that this figure is almost certainly incorrect. The errors in each of the five absorbance readings made at the calibration stage have been propagated into the calculated values of the slope and the intercept. The errors in a and b, together with an additional error in the absorbance reading of the unknown sample, have in turn been propagated into the predicted concentration (0.003 84 M). The use of statistical technique does not eliminate this error but it does help us to quantify our uncertainty. An approximate confidence interval for the true concentration of a sample which gives an absorbance reading equal to Y, is given by:

$$\left(\frac{Y-a}{b}\right) \pm \frac{t(\text{ESD})}{b} \sqrt{\left[1 + \frac{1}{n} + \left(\frac{Y-\bar{y}}{b}\right)^2 \Big/ Sxx\right]}$$

$$= \frac{(0.20 - 0.04)}{41.667} \pm \frac{3.18(0.027\,18)}{41.667} \sqrt{\left[1 + \frac{1}{5} + \left(\frac{0.20 - 0.29}{41.667}\right)^2 \Big/ 0.000\,066\right]}$$

$$= 0.003\,84 \pm 0.002\,074 \sqrt{(1 + 0.2 + 0.0707)}$$

$$= 0.003\,84 \pm 0.002\,074 \sqrt{(1.2707)} \tag{7.1}$$

$$= 0.003\,84 \pm 0.002\,34$$

(i.e. 0.001 50 M   to   0.006 18 M)

We can therefore be 95% confident that the true concentration of the unknown sample lies between 0.001 50 M and 0.006 18 M. Substituting several values of Y into the formula gives us the confidence intervals in Table 7.2. Note that these

Statistics for Analytical Chemists

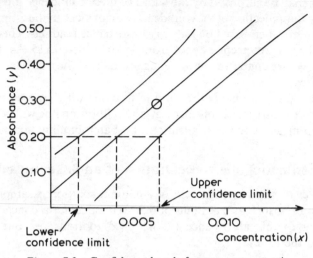

Figure 7.3 Confidence bands for true concentration

intervals get wider as the value of Y deviates further from $\bar{y}$ (0.29). This effect can be seen in Fig. 7.3 in which confidence bands have been drawn using the confidence intervals from Table 7.2.

Table 7.2 Confidence intervals for true concentration

Measured absorbance Y	95% confidence interval for true concentration			
0.00	$-0.000\,96 \pm 0.002\,88$	$-0.003\,84$ M	to	$0.001\,92$ M
0.10	$0.001\,44 \pm 0.002\,55$	$-0.001\,11$ M	to	$0.003\,99$ M
0.20	$0.003\,84 \pm 0.002\,34$	$0.001\,50$ M	to	$0.006\,18$ M
0.30	$0.006\,24 \pm 0.002\,27$	$0.003\,97$ M	to	$0.008\,51$ M
0.40	$0.008\,64 \pm 0.002\,37$	$0.006\,27$ M	to	$0.011\,01$ M
0.50	$0.011\,04 \pm 0.002\,61$	$0.008\,43$ M	to	$0.013\,65$ M

Returning to our unknown sample which gave an absorbance reading of 0.20; we can be 95% confident that the concentration of cuprammonium ion in this sample lies between 0.001 50 M and 0.006 18 M. Again this interval is disappointingly wide. The uncertainty in our estimates of the true slope, the true intercept and the error standard deviation gives rise to uncertainty in our estimate of the concentration of the unknown sample. How can this uncertainty be reduced and the confidence interval narrowed?

To answer this question let us return to the calculation of the confidence

interval. In the penultimate line of the calculation we have
0.003 84 ± 0.002 074√(1.2707). Obviously the width of the interval can only be
reduced if we reduce *either* the 0.002 074 *or* the 1.270 7 *or* both. We will consider
three ways of achieving our objective:

(a) Carry out a larger calibration experiment (i.e. using more than five known
     samples).
(b) Obtain a better estimate of the error standard deviation (e.g. from
     absorbance measurements in some other experiment).
(c) Carry out more than one determination on the unknown sample and use the
     mean of these determinations to estimate the concentration.

7.4 Increasing the size of the calibration experiment

We have already noted that the width of the confidence interval for true
concentration is greater for an absorbance reading that differs considerably from
the mean absorbance ($\bar{y}$) recorded during the calibration experiment. This results
in the curvature of the confidence bands in Fig. 7.3 and can be traced back to a
particular term in the formula, $(y - \bar{y})/b$. To remove this complication let us
consider only those unknown samples which give an absorbance reading equal to
$\bar{y}$. This will simplify the formula considerably to:

$$(y - a)/b \pm \frac{t(\text{ESD})}{b}\sqrt{\left(1 + \frac{1}{n}\right)}$$

The effect of increasing the size of the calibration experiment can now be easily
seen by substituting several values of n into this simplified formula. The
calculated confidence intervals are tabulated in Table 7.3 and illustrated
graphically in Fig. 7.4.

In Table 7.3 and Fig. 7.4 we can see the relationship between the size of the
calibration experiment and the width of the subsequent confidence interval for
the true concentration of an unknown sample. The smallest possible experiment,

Table 7.3 Width of confidence interval for true concentration

Number of samples in calibration experiment (n)	t value with ($n-2$) d.f.	Width of 95% confidence interval $\pm\dfrac{t(ESD)}{b}\sqrt{\left(1 + \dfrac{1}{n}\right)}$
3	12.71	±14.68 (ESD/b)
6	2.78	±3.00 (ESD/b)
12	2.23	±2.32 (ESD/b)
24	2.07	±2.12 (ESD/b)
48	2.01	±2.03 (ESD/b)
Infinity	1.96	±1.96 (ESD/b)

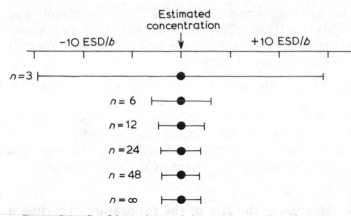

Figure 7.4 Confidence interval depends upon size of experiment

using only three known samples, gives rise to a very wide interval because the *t*-value with only one degree of freedom is very large. If the number of samples is increased to six the residual standard deviation then has four degrees of freedom and the *t*-value is reduced considerably. Clearly there is much to be gained by using six samples rather than three at the calibration stage, but there is little to be gained by a further increase to twelve samples.

It was asserted in the previous paragraph that the smallest possible calibration experiment requires the use of *three* samples. This is only true if the results of the experiment are to be used for the dual purpose of:

(a) Estimating the equation of the calibration line.
(b) Estimating the error standard deviation.

If the error standard deviation could be estimated by some other means then *two* samples would be the minimum requirement. Furthermore, if the analyst were prepared to assume that the true line passed through the origin, then the minimum number of samples would be further reduced to *one*. It is dangerous, however, to carry out the absolute minimum size of experiment because this leaves no opportunity to check the assumptions underlying the statistical analysis. We do not therefore advocate that the minimum size of calibration experiment should be used except in special circumstances. On the other hand, we would strongly recommend that the error standard deviation should be estimated before the calibration experiment is carried out.

7.5 An external estimate of error standard deviation

It was suggested in Chapter 6 that the analyst should explore the variability of a test method by means of a small-scale 'precision experiment' before carrying out any other experiment that will require statistical analysis of its results. The main purpose of this preliminary investigation is to explore the possibility that the precision of the test method may depend upon the level of concentration being measured. If we conclude that the random errors are 'relative' rather than 'fixed' then the statistical techniques can be appropriately modified. For example, the formulae we used to fit the calibration line would need to be modified if the error standard deviation were related to the concentration. These modifications will be discussed in the next chapter.

Suppose that a 'precision experiment' had been carried out and it revealed that the precision of the test method were *not* related to the concentration of the ammonia solution. Suppose, further, that this investigation produced an estimate of the error standard deviation equal to 0.027 18 based on ten degrees of freedom. (This is actually equal in value to the residual standard deviation calculated earlier in this chapter). Suppose, finally, that we were to carry out a calibration experiment using *only two* samples with concentrations of 0.002 M and 0.012 M, and these gave absorbance readings of 0.123 and 0.54, then we would obtain the same regression equation ($y = 0.04 + 41.66x$) as that fitted earlier.

Using this equation and the precision estimate we can now calculate a confidence interval for the true concentration of the unknown sample that gave an absorbance reading of 0.20. We will use the same formula that we used earlier, with $y = 0.20$, $a = 0.04$, $b = 41.67$ and $ESD = 0.027\,18$ being unchanged. The number of samples in the calibration is only two (i.e. $n = 2$), *but the t-value will have ten degrees of freedom* (i.e. $t = 2.23$) and $Sxx = 0.000\,05$.

The 95% confidence interval for the true concentration is given by:

$$\left(\frac{Y-a}{b}\right) \pm t(ESD)/b \sqrt{\left[1 + \frac{1}{n} + \left(\frac{Y-\bar{y}}{b}\right)^2 \bigg/ S_{xx}\right]}$$

$$= \left(\frac{0.20 - 0.04}{41.667}\right) \pm 2.23(0.027\,18)/41.667 \sqrt{\left[1 + \frac{1}{2} + \left(\frac{0.20 - 0.3315}{41.667}\right)^2 \bigg/ 0.000\,05\right]}$$

$$= 0.003\,84 \pm 0.001\,455\sqrt{(1 + 0.5 + 0.1992)}$$

$$= 0.003\,84 \pm 0.001\,455\sqrt{(1.6992)} \tag{7.2}$$

$$= 0.003\,84 \pm 0.001\,90$$

(i.e. 0.001 94 M   to   0.005 74 M)

We can be 95% confident that the true concentration of the unknown sample lies between 0.001 94 M and 0.005 74 M. This interval is narrower than that calculated earlier. Comparing Equation (7.2) above with Equation (7.1) in the earlier calculation we can see that the smaller *t*-value has given rise to a narrower interval despite an increase in the number under the square root sign. The latter has increased because we carried out a smaller calibration experiment using only two samples compared with the five samples in the earlier experiment. Let us now explore the third alternative strategy for reducing the width of the confidence interval.

7.6 Repeat determinations on the unknown sample

In Equation (7.2) we see that the largest of the three numbers under the square root sign is the '1'. This cannot be reduced by increasing the number of samples in the calibration experiment or by obtaining a better estimate of the error standard deviation. It can be shown, however, that this '1' appears in the confidence interval formula because of the random error in the absorbance reading of the unknown sample. If we were to make m determinations on the unknown sample the '1' in the formula would change to $1/m$.

Our unknown sample gave an absorbance reading of 0.20. Suppose this figure was actually the mean of *four* absorbance readings made on the unknown sample. (Based on four separate preparations, of course, not simply four readings of the instrument.) The predicted concentration would still be 0.003 84 M but we would surely have more faith in this prediction and we feel that a narrower confidence interval would be appropriate.

With m repeat determinations on the unknown sample the confidence interval for the true concentration is given by:

$$\left(\frac{Y-a}{b}\right) \pm t(\text{ESD})/b \sqrt{\left[\frac{1}{m}+\frac{1}{n}+\left(\frac{Y-\bar{y}}{b}\right)^2 \Big/ Sxx\right]}$$

where n is the number of samples used in the calibration)

$$= \frac{0.20-0.04}{41.667} \pm 3.18(0.027\,18)/41.667 \sqrt{\left[\frac{1}{4}+\frac{1}{5}+\left(\frac{0.20-0.29}{41.667}\right)^2 \Big/ 0.000\,066\right]}$$

$$= 0.003\,84 + 0.002\,074\sqrt{(0.25+0.2+0.0707)}$$

$$= 0.003\,84 \pm 0.002\,074\sqrt{(0.5207)}$$

$$= 0.003\,84 \pm 0.001\,50$$

(i.e. 0.002 34 M   to   0.005 34 M)

Obviously there is considerable scope for reducing the width of the confidence interval if the analyst is prepared to carry out repeat determinations on the unknown sample. If the repeat determinations are supported by a very good precision estimate for the test method, and a large-scale calibration experiment, considerable improvement in the confidence interval is possible. Clearly the benefits must be weighed against the effort involved.

7.7 Simulation of a calibration experiment

We have calculated a confidence interval for the true concentration of an unknown sample. The formula we used was rather complex, though it can be derived from the basic ideas of error propagation introduced in Chapter 3. Though we have explored different variants of the formula and discussed means of reducing the width of the confidence interval, we have only considered *one set of data.* The reader may suspect that the conclusions we have drawn are somewhat dependent upon the peculiarities of the data we have used. It would be reassuring, therefore, to take data from *several* calibration experiments and to calculate a confidence interval for each.

This objective can be achieved without too much effort if we use a computer to simulate a succession of calibration experiments. We must first specify the concentrations of the calibration samples. Let us use 0.002 M, 0.003 M, 0.005 M, 0.008 M and 0.012 M, which were used in the original experiment (see Table 7.1). Then we must specify the true equation and the distribution of the errors in the absorbance measurements. We will use the equation:

$$y = 0.05 + 42x$$

and specify that the errors should be normally distributed with a standard deviation of 0.03. (Such an arrangement could well have given rise to the data in Table 7.1.) Finally we must specify the true concentration of an unknown sample. We will use 0.007 M.

The computer program will now simulate a succession of calibration experiments. The results of the first five experiments are summarized in Table 7.4. In experiment no. 1 the absorbance measurements for the five calibration standard were:

| 0.137 | 0.223 | 0.299 | 0.386 | 0.574 |

These contain random errors that are based on random numbers generated in the computer. The absorbance measurements in the first experiment are therefore very unlikely to be repeated exactly in any later experiment. This point is confirmed if we examine the calculated slopes in Table 7.4. Each slope is different and they are scattered around the true slope of 40.0.

Table 7.4 Simulation of a calibration experiment

Experiment no.	Calculated slope (b)	Calculated intercept (a)	Calculated residual standard deviation	Absorbance of unknown sample	Predicted concentration	Confidence interval for true concentration
1	40.89	0.0774	0.0226	0.370	0.007 16	0.005 12 to 0.009 20
2	44.11	0.0711	0.0284	0.386	0.007 14	0.004 77 to 0.009 51
3	41.46	0.0565	0.009 34	0.378	0.007 76	0.006 92 to 0.008 60
4	44.33	0.0599	0.0178	0.329	0.006 07	0.004 60 to 0.007 54
5	38.30	0.0747	0.0191	0.337	0.006 86	0.005 01 to 0.008 70
True values	40.00	0.0500	0.0300		0.007 00	

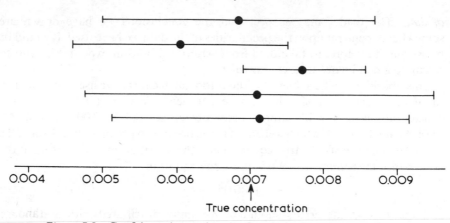

Figure 7.5   Confidence interval varies from experiment to experiment

The calculated intercepts also vary from experiment to experiment. It is simply a coincidence that, in the first five experiments, all the intercepts exceed the true intercept of 0.05. We also find variation in the other columns of Table 7.4 culminating in the widely differing confidence intervals that are presented graphically in Fig. 7.5.

We see that every one of the five confidence intervals in Fig. 7.5 does include the true concentration. If we were to repeat the calibration experiment many many times we would expect that approximately 5% of the confidence intervals would fail to include the true concentration. Requesting 100 more experiments we find that 7 (i.e. 7%) of the 100 confidence intervals do not include 0.007 M. A more ambitious run of the program reveals that, in 1000 experiments, only 43 (i.e. (4.3%) of the confidence intervals exclude the true value. This deviation from the expected 5% is no doubt partly due to sampling error but the confidence interval formula is only approximate and this further error would be expected to give a figure less than 5%.

The use of simulation can be reassuring and may well prove useful when we consider more complex calibration problems in the next chapter.

7.8   A calibration line through the origin

We remarked earlier that the least squares regression line fitted to the data in Table 7.1 does *not* pass through the origin and we wondered whether it would be better to fit an equation of the form $y = bx$. Several arguments can be put forward to support the fitting of a line which is constrained to pass through the origin:

(a) The Beer–Lambert Law, concerned with the physics underlying spectro-photometry, would suggest that the true relationship between absorbance and concentration is described by a straight line passing through the origin. (The Beer–Lambert Law could be expected to fall down at higher concentrations but these would be well above 0.012 M which was the maximum used in the calibration experiment.)

(b) If the true relationship *was* $y = \beta x$ and we fitted an equation of the form $y = ab + x$, then it is most unlikely that b would be equal to β or that a would be equal to zero, because of the random errors in the measured absorbance (y).

(c) In our fitted equation, $y = 0.04 + 41.667x$, a was *not* equal to zero, but a confidence interval for the true intercept *did* include zero.

There are good reasons therefore why it would be better to fit an equation of the form $y = bx$ than to fit the equation $y = 0.04 + 41.667x$. Later we will consider a disadvantage of fitting a line through the origin and then we will explore other alternatives; but first we will fit $y = bx$ and use this new equation to estimate the concentration of an unknown sample. Using the data in Table 7.1 and the formulae introduced in Chapter 5, we calculate:

$$\sum xy = 0.011\,45 \qquad \sum x^2 = 0.000\,246 \qquad \sum y^2 = 0.5373$$

$$\text{Slope } (b) = \sum xy / \sum x^2$$

$$= 0.011\,45/0.000\,246$$

$$= 46.5447$$

$$\text{Residual sum of squares} = \sum y^2 - b^2 \sum x^2$$

$$= 0.5373 - (46.5447)^2(0.000\,246)$$

$$= 0.004\,363$$

$$\text{Residual standard deviation} = \sqrt{(0.004\,363/4)}$$

$$= 0.033\,03$$

A confidence interval for the true slope is given by:

$$b \pm t(\text{ESD})/\sqrt{\sum x^2}$$

Again we will use the residual standard deviation to estimate the error standard derivation (ESD). This gives the following 95% confidence interval:

$$= 46.5447 \pm 2.78(0.033\,03)/\sqrt{(0.000\,246)}$$

$$= 46.5447 \pm 5.8544$$

$$\text{(i.e. } 40.69 \quad \text{to} \quad 52.40)$$

The equation of the best straight line through the origin is $y = 46.5x$ which is not very different from the best line through the centroid ($y = 0.04 + 41.667x$) over the range of concentrations used in the calibration. We note however that the confidence interval for the true slope extends from 40.69 to 52.40 for the line through the origin whereas the confidence interval for the true slope of the other line was much wider (31.03 to 52.31). By assuming that the true line must pass through the origin we get a better estimate of the true slope; provided of course that the assumption is justified.

The purpose of fitting the calibration line is, of course, to enable us to predict the concentration of an unknown sample. Can we expect better predictions of concentration from our new line than we got from the best line through the centroid? For an unknown sample that gives an absorbance reading equal to Y the predicted concentration is Y/b and a confidence interval for the true concentration of the sample is given by:

$$\frac{Y}{b} \pm \frac{t(\text{ESD})}{b} \sqrt{\left[1 + \left(\frac{Y}{b}\right)^2 \bigg/ \sum x^2\right]}$$

For the sample which gave an absorbance reading of 0.20, the 95% confidence interval is:

$$\frac{0.20}{46.54} \pm \frac{2.78(0.033\,03)}{46.54} \sqrt{\left[1 + \left(\frac{0.20}{46.54}\right)^2 \bigg/ 0.000\,246\right]}$$

$$= 0.004\,293 \pm 0.001\,973\sqrt{1 + 0.075\,07)}$$

$$= 0.004\,293 \pm 0.002\,046$$

$$(\text{i.e. } 0.002\,247 \text{ M} \quad \text{to} \quad 0.006\,339 \text{ M})$$

We are 95% confident that the true concentration of the unknown sample lies between 0.0022 M and 0.0063 M. This is a little narrower than the interval calculated after fitting $y = 0.04 + 41.667x$, which extended from 0.0015 M to 0.0062 M.

Substituting several values of Y into the confidence interval formula gives the intervals in Table 7.5. Note that these intervals get wider as the value of Y increases but this effect can barely be seen in Fig. 7.6 which portrays the confidence bands based on the intervals in Table 7.5.

Perhaps it would be timely to remind the reader of an assumption which applies throughout this chapter. We are assuming that the random errors of the

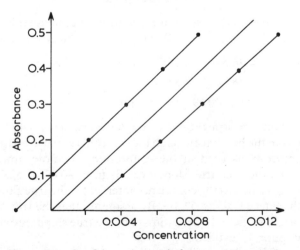

Figure 7.6 Confidence intervals for true concentration

Table 7.5 Confidence intervals for true concentration

Measured absorbance Y	95% confidence interval for true concentration		
0.00	0.00000 ± 0.00197	-0.00197 M to	0.00198 M
0.10	0.00215 ± 0.00199	0.00016 M to	0.00414 M
0.20	0.00430 ± 0.00205	0.00225 M to	0.00635 M
0.30	0.00645 ± 0.00213	0.00432 M to	0.00858 M
0.40	0.00859 ± 0.00225	0.00634 M to	0.01084 M
0.50	0.01074 ± 0.00239	0.00835 M to	0.01313 M

test method are *not* related to the concentration being measured. If the random errors were proportional to the concentration, for example, we would get confidence bands which differed greatly from those in Fig. 7.6, as we shall see in the next chapter.

It was pointed out earlier that one justification for fitting $y = bx$, rather than $y = a + bx$, is the physical law which relates absorbance and concentration. We should not forget, however, that the presence of a significant fixed bias would nullify this justification. Obviously a fixed bias would cause all the data points to be shifted upwards or downwards on the calibration graph and the true intercept would not be equal to zero. Bias is a constant problem and should certainly not be ignored at the calibration stage.

7.9   Correcting for bias in calibration

There are several ways in which we could attempt to estimate the effect of fixed bias at the calibration stage and to use this estimate to correct the calibration chart. We will describe one particular method which has two advantages in that it is simple to use and is already widely practised by analysts. To make use of this method of correction we would need to extend our calibration experiment by including an extra sample which contained *zero* concentration of the determinand. We will refer to this sample as a 'blank'. We would then subtract the absorbance of the blank from each of the other absorbance readings and fit a calibration line of the form $y = bx$ to the corrected data points.

Let us suppose that a blank had been included in the calibration experiment which gave rise to the data in Table 7.1. Let us further suppose that the blank had given an absorbance of 0.06. (This is perfectly consistent with the calculated intercept (0.04) and the estimated error standard deviation (0.027 18) that resulted from fitting $y = a + bx$ to the data in Table 7.1.) The extended data set is given in Table 7.6 and the corrected absorbance figures in Table 7.7.

In Table 7.7 the corrected absorbance is represented by z to distinguish it from the measured absorbance which was represented by y. Thus $z = y - y_0$ where y_0 is

Table 7.6   Calibration experiment extended by inclusion of a blank

Concentration (x)	0.000	0.002	0.003	0.005	0.008	0.012
Absorbance (y)	0.06	0.12	0.14	0.27	0.40	0.52

Table 7.7   Subtraction of blank response to correct for bias

Concentration (x)	0.002	0.003	0.005	0.008	0.012
Corrected absorbance (z)	0.06	0.08	0.21	0.34	0.46

the measured absorbance of the blank. We will now fit an equation of the form $z = bx$ to the data in Table 7.7. Calculation of the slope and of the residual standard deviation will be based on the usual formulae except that y is replaced by z:

$$\sum xz = 0.009\,65 \qquad \sum x^2 = 0.000\,246 \qquad \sum z^2 = 0.3813$$

$$\text{Slope } (b) = \sum xz / \sum x^2$$
$$= 0.009\,65/0.000\,246$$
$$= 39.2276$$

$$\text{Residual sum of squares} = \sum z^2 - b^2 \sum x^2$$
$$= 0.3813 - (39.2276)^2 (0.000\,246)$$
$$= 0.002\,754$$

$$\text{Residual standard deviation} = \sqrt{(0.002\,754}$$
$$= 0.026\,24$$

The calculations above have followed a well trodden path but, if we want to calculate confidence intervals for the true slope and the true concentration of an unknown sample, we must resort to the use of new formulae. This need arises because of the bias correction. It is unfortunate that the measured absorbance of the blank (0.06) contains a *random error*, in addition to the systematic error that we are seeking to correct. The result of correcting for bias, therefore, is to introduce extra random variation with a consequent loss of precision. The underlying mathematics becomes much more complex but an approximate confidence interval is given by:

Confidence interval for true slope is:

$$b \pm \frac{t(\text{ESD})}{\sqrt{\sum x^2}} \sqrt{\left[1 + \frac{(\sum x)^2}{\sum x^2}\right]}$$

A further problem introduced by the blank correction is that the residual standard deviation from the fitted line does *not* give an unbiased estimate of the error standard deviation (ESD). In fact the relationship between the residual standard deviation based on the z values and the error standard deviation of the x values is rather complicated. It can be shown however that 'the residual standard deviation divided by 1.1' gives a good estimate of the error standard deviation.

$$\text{Error standard deviation (ESD)} = \text{RSD}/1.1$$
$$= 0.026\,24/1.1$$
$$= 0.023\,85$$

Using this estimate we can now calculate a 95% confidence interval for the true slope:

$$39.23 \pm \frac{2.78(0.023\,85)}{\sqrt{(0.000\,246)}} \sqrt{\left[1 + \frac{(0.03)^2}{0.000\,246}\right]}$$
$$= 39.23 \pm 4.2276\sqrt{(4.6585)}$$
$$= 39.23 \pm 9.12$$
$$\text{(i.e. } 30.11 \quad \text{to} \quad 48.35)$$

This confidence interval for the true slope is similar in width to those we have seen earlier in this chapter. The uncertainty in the calculated slope will, of course, be propagated into the estimated concentration of any unknown sample. Before we calculate a confidence interval to quantify this uncertainty let us consider how the calibration line will be used.

By subtracting the measured absorbance of the blank we have, hopefully, removed from the calibration line the effect of any fixed bias in the test method. When we attempt to use the calibration line to estimate the concentration of an unknown sample we must take into account that the measured absorbance of this sample will also be subject to fixed bias. If the *same* bias were present during *both* calibration and operation of the test method, then there would be no need to correct for bias. It will often be the case, however, that calibration and operational use take place on different days or in different batches. If the fixed bias changes from day to day, or batch to batch, then the absorbance of the unknown sample will require a *different* correction to that used at the calibration stage. A suitable correction would be obtained by measuring the absorbance of another blank during the same day/batch as the unknown sample.

Let us suppose that a blank is measured immediately after the unknown sample and that the absorbance measurement of the blank is 0.01. This differs from the blank absorbance of 0.06 measured at the calibration stage. (The difference between the two is however less than 0.08, which is a rough estimate of the repeatability of the test method.) Subtracting the absorbance of the blank (0.01) from the absorbance of the unknown sample (0.20) gives a corrected absorbance of 0.19. Substituting $z = 0.19$ into our calibration equation ($z = 39.23x$) gives an estimated concentration of 0.0048 M. An approximate confidence interval for the true concentration of an unknown sample which has a corrected absorbance of Z, can be calculated from:

$$\frac{Z}{b} \pm \frac{t(\text{ESD})}{b} \sqrt{\left[2 + \frac{Z^2}{b^2 \sum x^2}\left(1 + \frac{(\sum x)^2}{\sum x^2}\right)\right]}$$

Substituting $Z = 0.19$, $b = 39.23$, $t = 2.78$, $\text{ESD} = 0.023\,85$, $\sum x^2 = 0.000\,246$ and $\sum x = 0.03$ gives the following 95% confidence interval:

$$\frac{0.19}{39.23} \pm \frac{2.78(0.023\,85)}{39.23} \sqrt{\left[2 + \frac{(0.19)^2}{(39.23)^2(0.000\,246)}\left(1 + \frac{(0.03)^2}{0.000\,246}\right)\right]}$$

$$= 0.0048 \pm 0.001\,690\sqrt{(2.442)}$$

$$= 0.0048 \pm 0.0062$$

(i.e. 0.0022 M to 0.0074 M)

How does this confidence interval compare with that calculated earlier when no bias correction was in use? On that occasion the estimated concentration was 0.0038 M and the interval extended from 0.0015 M to 0.0062 M. We see that the estimated concentration has changed because of the bias correction and we note that the interval has increased in width. In fact the difference in predicted concentration ($0.0048 - 0.0038 = 0.0010$ M) is rather small compared with the width of either confidence interval. If, however, the bias had been greater in either the calibration experiment or the operational use of the method, then the reduction in precision might well have been accompanied by an increase in accuracy. There are, of course, many ways in which precision can be improved as we saw earlier.

7.10 Summary

In this chapter we have looked at some of the statistical problems associated with calibration. We used the method of least squares to fit a calibration line through the centroid and a calibration line through the origin. In both cases we calculated a confidence interval for the true concentration of an unknown sample. Various strategies for reducing the width of this interval were considered and we saw the advantage of estimating the precision of the test method before attempting to calibrate. Finally we explored the use of blanks to correct for batch to batch changes in bias.

Throughout this chapter we have assumed that the random errors of the test method were *not* related to the concentration of the determinand. In other words we assumed a constant error standard deviation. In the next chapter we will reconsider the statistical problems of calibration when the error standard deviation is related to concentration.

Problems

(1) Phixerchem produce photographic grade potassium thiocyanate to a specification laid down in BS 3833. One criterion set by the British Standard

is that the lead content should not exceed 0.1% (1000 p.p.m.). The Chief Chemist of Phixerchem is anxious to replace the present laborious test by an automatic one in which a microamp reading is obtained from each sample. Although the chemist expects the method to be less accurate than the one given in the British Standard, he claims that the accuracy of the new method is more than satisfactory since the lead content is usually well below the specified maximum. To obtain a calibration curve and an indication of its precision the chemist designed an experiment covering concentrations up to 1000 p.p.m. (Higher concentrations can be analysed by using an appropriate dilution of the test solution.) The results of the calibration experiment are given below.

Concentration in p.p.m. (x)	100	200	300	400	500	600	700	800	900	1000
Microamps (y)	62	107	136	177	230	254	296	347	390	421

Summary statistics: $n = 10$; $\bar{x} = 550$; $Sxx = 825\,000$; $\bar{y} = 242$; $Syy = 134\,260$; $Sxy = 332\,400$.

(a) Using the summary statistics calculate the slope and intercept of the calibration line.
(b) Calculate the residual sum of squares using the formula:
$$\text{Residual sum of squares} = Syy - b^2 Sxx$$
(c) Calculate the residual standard deviation.
(d) Calculate a 95% confidence interval for the true concentration when a single determination gives a microamp reading of 400. (Use the residual standard deviation from (c) as an estimate of ESD.)
(e) On the basis of the confidence interval calculated in (d), what is the maximum microamp reading which will give a lead content that satisfies BS 3833?

(2) There is evidence that the method includes a fixed bias which varies from batch to batch and therefore the standard procedure states the reading of a blank should be subtracted from each reading. In the calibration experiment the blank had a reading of 20.0. Subtracting this reading gives the following corrected results:

Concentration in p.p.m. (x)	100	200	300	400	500	600	700	800	900	1000
Corrected microamps (z)	42	87	116	157	210	234	276	327	370	401

Summary   statistics: $\sum x = 5500$; $\sum x^2 = 3\,850\,000$; $\sum xz = 1\,553\,400$; $\sum z^2 = 627\,100$.

(a) Fit a regression line through the origin.

(b) Calculate the residual standard deviation and estimate the error standard deviation.

(c) Calculate a 95% confidence interval for the true concentration of a sample that gives a microamp reading of 420 whilst a blank in the same batch gives a reading of 40 microamps.

(d) Comparing the confidence intervals in Problem (1) and (2) we notice that subtraction of a blank has increased the width of the interval. What benefit has been gained by the subtraction of a blank to compensate for the loss in precision?

(3) After examination of a number of calibration experiments, Phixerchem decide that the slope varies significantly from experiment to experiment but the RSD is fairly constant at a value of 5.50 p.p.m. To overcome the variability in slope they decide to calibrate each batch by including three standards. These are:

(i) A blank;

(ii) A standard at 1000 p.p.m. This standard is used to calculate the slope using the formula:

$$\frac{\text{standard reading} - \text{blank reading}}{1000}$$

(iii) A standard at 500 p.p.m. This standard is included as a check that certain assumptions – for example, linearity, absence of outliers – have not been violated. The reading obtained with this standard should, on conversion, give a 95% confidence interval which includes the true concentration.

The three standards in a batch of samples gave the following microamp readings:

Blank   18
1000 p.p.m.   429
500 p.p.m.   217

(a) Calculate the slope.

(b) Remembering that only one corrected determination (z) is used to estimate the slope, calculate:

(i) $\sum z$;
(ii) $\sum z^2$.

(c) Using an ESD of 5.5 p.p.m. based on infinite degrees of freedom, calculate a 95% confidence interval for the true concentration when a determination gives a reading of 217 and a blank of 18. Does the concentration of 500 p.p.m. for the standard fall within the confidence interval? Decide if there is any evidence that the assumptions pertaining to the calibration have been violated.

(4) Dr A. Smith, an Analytical Consultant has been called to the Rubovian Embassy in Bradfox to determine the antimony content of a top secret sample. The Ambassador is reluctant to divulge the source of the sample but the Analyst has managed to discover that it is in the form of a dark red viscous fluid which is susceptible to analysis by HPLX (a method recently invented by Smith). He is also assured that the concentration of antimony can be expected to be between 0 and 100 p.p.m.

On arrival at the Embassy he learns that the result is required very urgently and that there is sufficient reagent for only ten determinations. Which of the following strategies can be expected to give Smith the narrowest confidence interval for the true concentration of antimony in the unknown sample if the true concentration is very close to 50 p.p.m.? (Do not assume that the calibration line passes through the origin.)

(a) Use five standards at 50 p.p.m. to estimate the error standard deviation then use four standards at 0, 33, 66 and 100 p.p.m. in a calibration experiment and finally make one determination on the unknown sample.

(b) Use nine standards at 0, 12.5, 25, ..., 100 p.p.m. in a calibration experiment and then make one determination on the unknown sample.

(c) Use six standards at 0, 20, 40, 60, 80 and 100 p.p.m. in a calibration experiment and then make four determinations on the unknown sample.

(d) Use three standards at 0, 50 and 100 p.p.m. in a calibration experiment and then make seven determinations on the unknown sample.

8

Calibration – further difficulties

8.1 Introduction

In the previous chapter we discussed the fitting of a straight line to the data from a calibration experiment. The equation of the line was used to predict the true concentration of an unknown sample and the precision of this prediction was quantified by calculating a confidence interval. We focused attention on this confidence interval in order to explore various ways in which the *precision* of determinations might be improved. We then turned our attention to *bias* in the test method and considered how this might be corrected by the use of blanks.

Throughout the previous chapter we assumed that the precision of the test method was the same at all levels of concentration throughout a particular range. It is necessary that this should be so if we are to use the method of *ordinary least squares* or *ordinary regression* to fit the calibration line.

Obviously there are many analytical methods which do *not* have a precision that is constant over a wide range. In such cases we must modify our approach to make use of *weighted regression*.

8.2 Weighted regression

With several sets of data we have fitted what we described as *the best straight line*. In doing so we used the method of least squares. This method gives us a straight line that has a smaller residual sum of squares than any other line would have.

When using the method of least squares we attached equal importance to every point on the graph. This is perfectly reasonable if we know nothing about the errors in the data. If, however, we know, in advance, that point A is more likely to have a larger error than is point B, then it may be wise to attach more importance to B and less to A. We could do so by using *weighted least squares*, or *weighted regression* as it is often called. With this method we give a different weight to each point and we minimize the *weighted residual sum-of-squares*.

The extent to which the weighted regression line differs from the ordinary regression line will depend on how we allocate weights to the points. If we give *equal* weight to each point then the two methods will give the same line. When allocating weights it is usual to give smaller weights to those points that are likely to have larger errors and to give large weight to those points that are likely to have smaller errors.

But how are we to know what errors we are likely to get in a calibration experiment?

It was suggested in Chapter 6 that we should investigate the 'error structure' of a test method *before* carrying out a calibration experiment. In that chapter we attempted to answer the question 'Is the precision of the test method related to the concentration of the determinand?' We explored a set of data that resulted from an analyst making six repeat determinations at each of five concentrations. We found some evidence that the variability of the method increased with increasing concentration, but this evidence was not conclusive. Had we been able to prove, beyond reasonable doubt, that the precision *was* related to the concentration then it would have been unwise to use the ordinary least-squares method at the calibration stage.

In Chapter 7 we fitted a calibration line using the ordinary method of least squares. Throughout Chapter 7 we worked on the *assumption* that the error standard deviation was constant. Had an attempt been made to investigate the relationship between precision and concentration we might have found that the error standard deviation increased with concentration. Let us suppose that such an investigation *had* been carried out, before the calibration experiment, and that the report read as follows:

> The investigation was based on many determinations, made by several operators, in a large number of laboratories and the results clearly show that the variability of the test method is greater at higher concentrations of cuprammonium ion. The relationship between precision and concentration is quantified by the equation:

$$(\text{Standard deviation of absorbance measurements}) = 0.011$$
$$+ 3.81(\text{Concentration of cuprammonium ion}) \qquad (8.1)$$

In the light of this new evidence we will return to the data in Table 7.1 and fit a calibration line using weighted regression. The weights that we use will be related to the standard deviations given by equation (8.1). These are listed in Table 8.1, together with the data. The standard deviations have been used to calculate the 'one-sigma limits' portrayed in Fig. 8.1.

Table 8.1 Calibration data and predicted standard deviations

Concentration (M)	x	0.002	0.003	0.005	0.008	0.012
Absorbance	y	0.12	0.14	0.27	0.40	0.52
Predicted SD	σ	0.0186	0.0224	0.0301	0.0415	0.0567

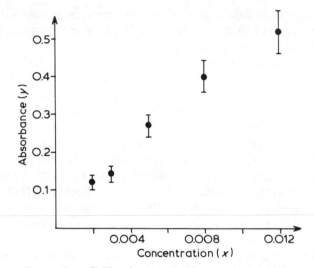

Figure 8.1   Calibration data with 'one-sigma limits'

The 'one-sigma limits' Fig. 8.1 are simply $(y+\sigma)$ and $(y-\sigma)$. They indicate a band, in which we might reasonably expect the point to lie, if the experiment were to be repeated. The clear indication from Fig. 8.1 is that we can place more confidence in the points with lower concentration. Weighted regression will take this into account.

The weight to be given to each point is calculated in Table 8.2. The standard deviations in the third column are identical to those in Table 8.1 and were calculated using Equation (8.1). From the standard deviations we can calculate the reciprocals of the variances in the fourth column. Dividing each entry in the fourth column by the mean of the fourth column (1374.28) gives the weights in the fifth column.

Table 8.2   Calculation of weights

	Concentration x	Absorbance y	Standard deviation σ	Reciprocal of variance $1/\sigma^2$	Weight w
	0.002	0.12	0.018 62	2884.3	2.099
	0.003	0.14	0.022 43	1987.7	1.440
	0.005	0.27	0.030 05	1107.4	0.806
	0.008	0.40	0.041 48	581.2	0.423
	0.012	0.52	0.056 72	310.8	0.226
Total	0.030	1.45		6871.4	5.000
Mean	0.006	0.29		1374.28	1.000

Note that the weights in Table 8.2 add up to 5.000. It is no coincidence that the sum of the weights is equal to the number of points. We have arranged that this will be so in order to simplify the regression formulae that we will use later. We could have used the entries in the fourth column of Table 8.2 as weights, but the calculations would have been rather messy and the formulae less meaningful.

Those points with weights greater than 1.000 will have more influence on the slope of the fitted line than they would have had if we used ordinary regression; whilst those points with weights less than 1.000 will have less influence. We see that the ratio of the largest weight to the smallest weight is 2.099/0.226 which is 9.3; thus the point with lowest concentration will have considerably more influence than the point with largest concentration.

The reader may wonder why each point should be given a weight that is proportional to the reciprocal of its error variance. The reason will become clear if we explore the method of 'weighted least squares' in which we choose a regression equation $y = a + bx$ so as to minimize the weighted residual sum of squares:

$$\text{Weighted residual sum of squares} = \sum[(\text{weight})(\text{residual})^2] \qquad (8.2)$$

We have seen in earlier chapters that a residual is largely a reflection of the random error in the absorbance measurement (provided, of course, that the true relationship between x and y really *is* linear). If we have larger errors in those points which have a high concentration, then these points would unduly influence the position of the calibration line if we used the ordinary least-squares method. In order to overcome this difficulty we could standardize the residuals such that:

$$\text{Standardized residual} = \left(\frac{\text{residual}}{\text{standard deviation from Equation (8.1)}}\right)$$

We could expect the standardized residuals to be roughly equal in magnitude regardless of concentration. Squaring and summing the standardized residuals would give the standardized residual sum of squares:

$$\text{Standardized residual sum of squares} = \sum\left(\frac{\text{residual}}{\text{standard deviation}}\right)^2 \qquad (8.3)$$

Comparing Equations (8.2) and (8.3) we see that we can achieve our objective by choosing our weights (w) so that:

$$w = 1/\sigma^2$$

We could, of course, choose our weights in some other way; but this approach does seem to be entirely reasonable and is certainly very often used. Regardless of how the weights are chosen, the slope and intercept of the weighted regression line are given by:

$$\text{Slope } (b) = Swxy/Swxx$$
$$\text{Intercept} = \bar{y}_w - b\bar{x}_w$$

where $\bar{y}_w = \sum wy/n$ and $\bar{x}_w = \sum wx/n$

and $Swxy = \sum w(x - \bar{x})(y - \bar{y})$ or $\sum wxy - n\bar{x}_w\bar{y}_w$

and $Swxx = \sum w(x - \bar{x})^2$ or $\sum wx^2 - n\bar{x}_w^2$

Table 8.3   Fitting a weighted regression line

x	y	w	wx	wy	wxy	wx^2	wy^2
0.002	0.12	2.099	0.004 198	0.251 88	0.000 503 76	0.000 008 396	0.030 225 6
0.003	0.14	1.466	0.004 338	0.202 44	0.000 607 32	0.000 013 014	0.028 341 6
0.005	0.27	0.806	0.004 030	0.217 62	0.001 088 10	0.000 020 150	0.058 757 4
0.008	0.40	0.423	0.003 384	0.169 20	0.001 353 60	0.000 027 072	0.067 680 0
0.012	0.52	0.226	0.002 712	0.117 52	0.001 410 24	0.000 032 544	0.061 110 4
Total 0.030	1.45	5.000	0.018 662	0.958 66	0.004 963 02 ($\sum wxy$)	0.000 101 176 ($\sum wx^2$)	0.246 115 ($\sum wy^2$)
Mean 0.006 ($\bar{x}$)	0.29 ($\bar{y}$)	1.000 ($\bar{w}$)	0.003 732 4 ($\bar{x}_w$)	0.191 732 ($\bar{y}_w$)			

$$Swxy = \sum wxy - n\bar{x}_w\bar{y}_w$$
$$= 0.004\,963\,02 - 5(0.003\,732\,4)(0.191\,732)$$
$$= 0.001\,384\,92$$
$$Swxx = \sum wx^2 - n(\bar{x}_w)^2$$
$$= 0.000\,101\,176 - 5(0.003\,732\,4)^2$$
$$= 0.000\,031\,522\,0$$
$$\text{Slope } (b) = Swxy/Swxx$$
$$= 0.001\,384\,92/0.000\,031\,522\,0$$
$$= 43.935$$
$$\text{Intercept } (a) = \bar{y}_w - b\bar{x}_w$$
$$= 0.191\,732 - 43.935(0.003\,732\,4)$$
$$= 0.0277$$

The equation of the weighted regression line is therefore $y = 0.028 + 43.935x$. This line is represented graphically in Fig. 8.2. A comparison of Fig. 8.2 with Fig. 7.2 in Chapter 7 reveals that the weighted regression line differs very little from the ordinary regression line, which had the equation $y = 0.040 + 41.667x$.

The reader might wonder whether we will *always* get similar equations by the two methods, ordinary regression and weighted regression. This point will be clarified when we use a simulation program to generate many sets of calibration data and use both techniques to fit a calibration line. Before we return to our

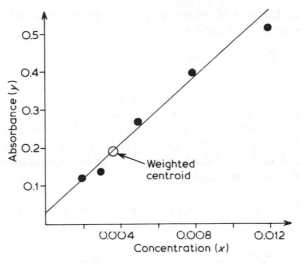

Figure 8.2 Calibration line by weighted regression

simulation program let us consider a confidence interval for the true concentration of an unknown sample. The width of this confidence interval can be very different when we use weighted regression.

8.3 Confidence interval with weighted regression

In Chapter 7 we calculated a confidence interval for the true concentration of an unknown sample using the formula:

$$\left(\frac{Y-a}{b}\right) \pm \frac{t(\text{ESD})}{b} \sqrt{\left[1 + \frac{1}{n} + \left(\frac{Y-\bar{y}}{b}\right)^2 \Big/ \text{Sxx}\right]}$$

where Y is the absorbance of the unknown sample. We used several values of Y to obtain the confidence bands plotted in Fig. 7.3. The reader may recall that the confidence interval was narrowest for an absorbance equal to the mean absorbance ($\bar{y}$), with the interval widening as Y deviated further from $\bar{y}$.

On that occasion we were calibrating a test method that was thought to have a constant error standard deviation. In this chapter we are assuming that the error standard deviation increases with the concentration:

$$\text{error SD} = 0.011 + 3.81(\text{concentration})$$

It is reasonable to expect a narrow confidence interval for a sample with low concentration and a wider confidence interval for a sample with higher concentration. This is indeed what we get if we use weighted regression and the weighted confidence interval formula:

$$\left(\frac{Y-a}{b}\right) \pm \frac{t(\text{ESD})}{b} \sqrt{\left[\frac{1}{w} + \frac{1}{n} + \left(\frac{Y-\bar{y}_w}{b}\right)^2 \Big/ Swxx\right]}$$

This formula is very similar to the one used in Chapter 7. The important difference between the two being the '$1/w$' having replaced the '1' under the square root. Thus samples with low concentration, having low error standard deviation and hence high weight, will give rise to narrow confidence intervals. This point should be clear in Table 8.4 which contains confidence intervals for the true concentration of several samples.

Table 8.4　Confidence intervals with weighted regression

Measured absorbance Y	Weight w	95% confidence interval for true concentration		
0.00	4.268	-0.00064 ± 0.00163	-0.00227	to	0.00099
0.10	1.685	0.00164 ± 0.00154	0.00010	to	0.00318
0.20	0.901	0.00392 ± 0.00183	0.00209	to	0.00575
0.30	0.560	0.00619 ± 0.00236	0.00383	to	0.00855
0.40	0.381	0.00847 ± 0.00301	0.00546	to	0.01148
0.50	0.276	0.01074 ± 0.00371	0.00703	to	0.01445

In the calculation of these confidence intervals we have used $t_3 = 3.18$, $a = 0.028$, $b = 43.935$, $\bar{y}_w = 0.191\,732$ and $Swxx = 0.000\,031\,5220$. The residual standard deviation (RSD) was used as an estimate of the error standard deviation and was calculated as follows:

Residual standard deviation $= \sqrt{[(Swyy - b^2\, Swxx)/(n-2)]}$

$\qquad\qquad\qquad\qquad\qquad = \sqrt{[(0.062\,2595 - 43.935^2(0.000\,031\,5220))/3]}$

$\qquad\qquad\qquad\qquad\qquad = 0.022\,077$

Comparing the confidence intervals in Table 8.4 with those in Table 7.2 we see that the weighted regression has given us much more realistic intervals which reflect the known error structure (i.e. $SD = 0.011 + 3.81x$). The confidence bands are plotted in Fig. 8.3.

8.4　Simulation of a calibration experiment

When using weighted regression we obtained a calibration line that differed very little from that fitted in the previous chapter using ordinary regression. One wonders if the two methods will give a similar line with *any* set of data. By means

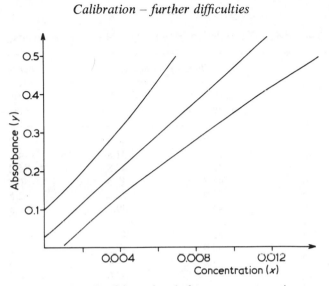

Figure 8.3   Confidence bands for true concentration

of the simulation program used in Chapter 7 we could repeat the calibration
experiment many times and, with each set of data we could fit *two* calibration
lines – one by ordinary regression and one by weighted regression. The program
was instructed to simulate 1000 calibration experiments using the true equation
($y=0.05+42.0x$) that we used in Chapter 7. Unlike the previous simulation,
which used a constant error standard deviation of 0.03, the program was
instructed to use $SD=0.011+3.81x$. The slopes of the fitted lines in the first 20
experiments are tabulated in Table 8.5.

Throughout the simulation the true equation was $y=0.05+42.0x$. We would
therefore have expected the mean slope to be 42.0 if we had continued indefinitely.
In some experiments (e.g. no. 2 and no. 16) the calculated slope differs
considerably from the true slope; but this is true for *both* methods. Amongst these
twenty experiments, any set of results that gives a high (low) slope by ordinary
regression also gives a high (low) slope by weighted regression. This point is
clearly illustrated in Fig. 8.4 which is based on the slopes from forty experiments.

In Fig. 8.4 the axes have been drawn at $x=42$ and $y=42$ because the true slope
in the simulation model was 42.0. A point at the intersection of the two axes
would represent a calibration experiment in which both methods gave a
calculated slope of 42.0. It would be very reassuring to find all 40 points tightly
clustered around this intersection. The broken lines in Fig. 8.4 divide the graph
into four quadrants and the number of points in each quadrant is given by the
circled number. In the two quadrants that contain 14 and 13 points weighted
regression has given a calculated slope closer to the true slope than has ordinary
regression. In the other two quadrants ordinary regression has got closer to the
truth. Thus weighted regression has proved superior in 27 of the 40 experiments.

Table 8.5 A simulation of twenty calibration experiments

Experiment no.	Slope of fitted line		Difference between the two slopes
	Ordinary regression	Weighted regression	
1	42.0	46.2	4.2
2	36.3	31.8	−4.5
3	42.6	47.8	5.2
4	46.7	47.5	0.8
5	52.9	50.4	−2.5
6	41.5	39.2	−2.3
7	37.7	40.7	3.0
8	42.4	42.2	−0.2
9	44.0	47.2	3.2
10	37.2	34.4	−2.8
11	48.3	48.6	0.3
12	38.0	40.6	2.6
13	38.5	42.8	4.3
14	38.6	37.3	−1.3
15	37.1	37.4	0.3
16	52.6	53.4	0.8
17	35.7	34.5	−1.2
18	43.6	46.1	2.5
19	40.6	39.5	−1.1
20	43.1	43.8	0.7
Mean	41.97	42.57	0.60
SD	5.031	5.831	2.657

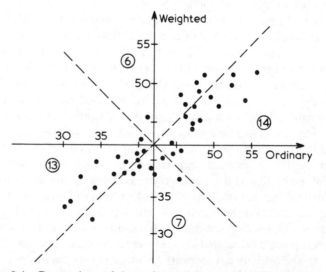

Figure 8.4 Comparison of slopes by ordinary and weighted regression

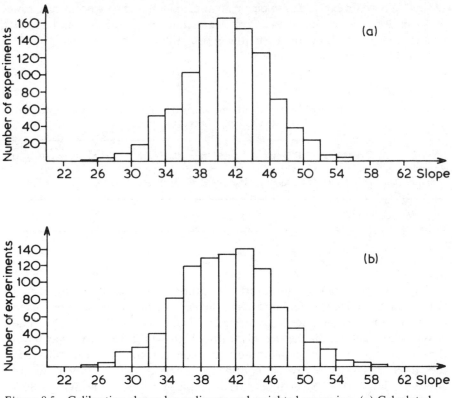

Figure 8.5 Calibration slopes by ordinary and weighted regression. (a) Calculated slopes by weighted regression, mean = 42.03, SD = 4.865; (b) calculated slopes by ordinary regression, mean 42.16, SD = 5.603

The reader may feel that neither method is very satisfactory because of the variation in slope from experiment to experiment. This variability is perhaps better seen in Fig. 8.5 which allows each of the regression methods to be examined in isolation. The two histograms in Fig. 8.5 summarize the results of all 1000 experiments and we can see that weighted regression gives a smaller spread of slopes; 95% of the weighted slopes lie in the range 32.4 to 51.5, whilst 95% of ordinary slopes lie within the range 30.8 to 53.6. Figure 8.5 supports the conclusion suggested by Fig. 8.4, that weighted regression is marginally superior to ordinary regression when the error standard deviation increases with concentration. To appreciate the *major* advantage of using weighted regression, however, we must examine the confidence intervals for the true concentration of an unknown sample. The confidence intervals given by the two methods are summarized in Table 8.6.

Table 8.6 tells us the number of experiments, out of 1000, in which the calculated 95% confidence interval did *not* contain the true concentration of the sample. We would expect this to occur on approximately 5% of occasions, which is 50 in 1000. When using *weighted* regression we get 38, 43 and 39 such occurrences. All three figures are below expectation but in accord with the 43% obtained by simulation in Chapter 7.

Table 8.6 The number of experiments in which the confidence interval does not include the true concentration

Number of experiments simulated	True conc. = 0.002 M		True conc. = 0.007 M		True conc. = 0.012 M	
	Weighted	Ordinary	Weighted	Ordinary	Weighted	Ordinary
1000	38	21	43	72	39	135

When we turn our attention to the calibration lines fitted by *ordinary* regression we find very different results. The number of confidence intervals that do not include the true concentration are 21, 72 and 135. Obviously the first figure is well below expectation, the second is rather high, whilst the third is way beyond the limits of sampling variation. At the higher concentration 865 (i.e. 1000–135) of the 1000 experiments gave a confidence interval that included the true concentration of 0.012 M. It appears that ordinary regression is giving a 86.5% confidence interval at this concentration, rather than the 95% that was asked for. Obviously we cannot trust the confidence intervals from ordinary regression when the precision of the test method is related to the concentration of the sample.

8.5   Calibration curves

Whilst discussing calibration we have confined our attention to the fitting of straight lines. We have used ordinary regression and we have used weighted regression, but with both methods we assumed that the true relationship between concentration and absorbance was linear. We had sound reasons for doing so:

(a) The Beer–Lambert Law suggests that we should expect a linear relationship at low concentrations of cuprammonium ion.
(b) Even if the relationship was non-linear at higher concentrations, we could avoid this region by diluting an unknown sample until its concentration fell within the range of our existing calibration (i.e. 0.002 M to 0.912 M).
(c) Fitting and using a straight line is very much easier than fitting and using a curve.

Will all analysts be as fortunate as our Chief Chemist? Is it *always* possible to dilute a sample to reduce the concentration of the determinand to within a known range? Do situations arise in which the relationship between concentration and measurement is *essentially* non-linear even at the lowest concentrations?

Undoubtedly there are *some* situations in which the analyst has to contend with the extra complexity of *curved* calibration. These would appear to be rather rare, however, and we do not intend to devote to non-linear calibration the amount of space that would be needed to deal with the problem satisfactorily. We will confine our discussion to just one set of data, with the main purpose of highlighting the dangers and the difficulties.

The data we have used in this chapter and in the previous chapter came from a calibration experiment in which the concentrations of the prepared samples ranged from 0.002 M to 0.012 M. The Chief Chemist of Indichem, who designed the experiment, now wishes to extend the investigation to cover higher concentrations of cuprammonium. It is understandable that he should want a calibration chart which covers the whole range of concentrations that are likely to be encountered in his laboratories.

The Chief Chemist considers that concentrations as high as 0.08 M could arise and he intends to supplement the data from the first experiment by measuring the absorbance of prepared samples which have concentrations of 0.016 M, 0.022 M, 0.032 M, 0.045 M, 0.064 M and 0.080 M. This will extend the range of the calibration chart considerably.

In the new experiment three measurements of absorbance are obtained at each of the six levels of concentration which gives the data in Table 8.7. The inclusion of repeat determinations at each concentration will allow the Chief Chemist to investigate the error structure before fitting a calibration curve.

Table 8.7   A wider range experiment

Concentration	0.016 M	0.022 M	0.032 M	0.045 M	0.064 M	0.080 M
Absorbance	0.78	1.05	1.35	1.47	1.71	1.83
	1.32	0.98	1.30	1.50	1.60	1.73
	0.81	1.00	1.40	1.55	1.67	1.69

A cursory inspection of Table 8.7 should lead to the realization that all is not well. The absorbance of 1.32, recorded for one of the three samples which had a concentration of 0.016 M, does not fit in with the rest of the data. An absorbance of such magnitude would be more at home in the third column of the table and one is tempted to speculate that a mistake was made in the preparation of this particular sample. Whatever the explanation for this obvious outlier the best course of action is to reject the absorbance measurement completely. The wisdom

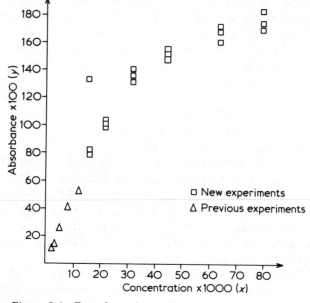

Figure 8.6   Data from the two calibration experiments

of this decision is confirmed by Fig. 8.6 which is a graph of the data in Table 8.7.

Further inspection of Table 8.17 or Fig. 8.6 might lead one to the less obvious, but equally important, realization that *the variability of the absorbance measurements is greater at higher concentrations.* This will be seen more clearly in the bottom rows of Table 8.8.

Table 8.8   Investigation of error structure

Concentration (x)	0.016 M	0.022 M	0.032 M	0.045 M	0.064 M	0.080 M
Absorbance (y)	0.78	1.05	1.35	1.47	1.71	1.82
	—	0.98	1.30	1.50	1.60	1.73
	0.81	1.00	1.40	1.55	1.67	1.69
Mean ($\bar{y}$)	0.795	1.010	1.350	1.507	1.667	1.747
Standard deviation (s)	0.0212	0.0361	0.050	0.0404	0.0557	0.0666

In Table 8.8 the mean and standard deviation of the y values has been calculated at each of the six levels of concentration. We see that the larger standard deviations tend to be associated with those concentrations which gave a high mean absorbance. This relationship is perhaps more noticeable in Fig. 8.7.

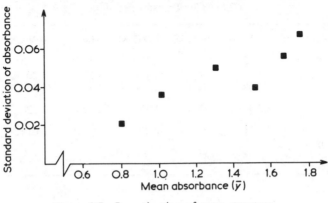

Figure 8.7 Investigation of error structure

Fitting the best straight line to the points in Fig. 8.7 we get the equation:

$$SD = -0.0737 + 0.0389 \text{ (mean)}$$

A confidence interval for the true slope does not include zero and the correlation coefficient (0.9165) is statistically significant. We must conclude, therefore, that the precision of the test method is inferior at higher concentrations.

So we have *two* problems. Figure 8.6 indicates that we need to fit a curve whilst Fig. 8.7 indicates that we need to take account of the variable precision of the test method whilst fitting the curve. We have already seen in this chapter how weighted regression can be used to overcome the second problem and curve-fitting programs are available to help us solve the first problem. Unfortunately the analyst is very unlikely to solve both problems simultaneously because he is unlikely to find a curve-fitting program that will allow weighting of the points. The analyst could write his own computer program, but he should be warned that curve-fitting poses certain technical problems that can only be overcome by the use of 'orthogonal polynomials'. Many available curve-fitting programs make use of orthogonal polynomials. Whilst these programs are rather complex and will not be described in this text, they are very easy to use as we shall see later. They are, however, intended to be used with homoscedastic data and we must eliminate the heteroscedascity in our data *before* we attempt to fit curves.

Fortunately, it is often possible to obtain a constant standard deviation by *transforming* the data. We will have more to say about data transformation when we discuss BS 5497 and IS 4259 in Chapter 12; for our present purposes we will simply assert that a suitable transformation of the data from our new experiment is:

$$Y = \log(10y)$$

The absorbance measurements arc multiplied by 10 before taking logs simply to avoid negative values of Y. The transformed data is shown in Table 8.9.

Statistics for Analytical Chemists

Table 8.9 Transformed data

Concentration (x)	0.016 M	0.022 M	0.032 M	0.045 M	0.064 M	0.080 M
Log (absorbance $\times 10$) (Y)	0.892	1.021	1.130	1.167	1.233	1.260
		0.991	1.114	1.176	1.204	1.238
	0.908	1.000	1.146	1.190	1.223	1.228
Mean of Y	0.900	1.004	1.130	1.178	1.220	1.242
Standard deviation of Y	0.011	0.015	0.016	0.012	0.015	0.016

The effectiveness of the transformation can be judged by inspecting Fig. 8.8 in which the variability of log absorbance is plotted against mean log absorbance for the six levels of concentration. With so few points, it is of course possible to see almost any pattern that you want to see, but a comparison of Fig. 8.8 with Fig. 8.7 does indicate that our transformation has been successful in removing the heteroscedasticity in the data. It is reasonable to suggest that the transformed data satisfies the homoscedasticity assumption.

Before we use regression analysis to fit an equation to the transformed data one or two points need to be noted concerning the use of transformations. Satisfying the homoscedasticity assumptions is not the *only* reason for transforming data. Quite often data is transformed in order to change a non-linear relationship into a linear relationship. Taking logs of x or y or both is often found to be useful when we are confronted with a relationship such as:

$$y = A e^{-Bx}$$
$$\text{or} \quad y = A x^B$$
$$\text{or} \quad y = A(B^x)$$

It cannot be stressed too strongly that the purpose of the transformation carried out in Table 8.10 was *not* to linearize the relationship between x and y. An examination of Fig. 8.9 should make this point clear. The data points in Fig. 8.9 are crying out for a curve not a straight line. Far from linearizing the relationship, the use of the transformation appears to have *increased* the curvature as we can see by comparing Fig. 8.9 with Fig. 8.6.

Let us now fit a curve to the data in Fig. 8.9. We will make use of a computer program which fits polynomial equations and prints excellent graphs. As we fit successively longer polynomial equations we will at each stage examine the corresponding graph to check the suitability of the equation for our calibration purposes. This is exactly what one would do in practice, if one were using the desk-top computer on which the program was run.

Figure 8.10 portrays the best linear equation. Clearly this equation is totally

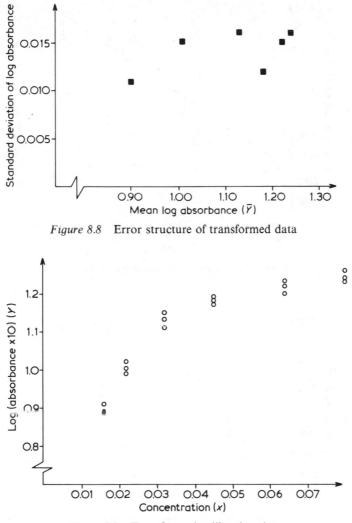

Figure 8.8 Error structure of transformed data

Figure 8.9 Transformed calibration data

inadequate for calibration purposes. There is no need to calculate confidence intervals to realize that predictions from this equation would be useless.

Figure 8.11 is a graphical representation of the best quadratic equation. Whilst this offers a big improvement in percentage fit over the linear equation (95.4% compared with 79.2%) it must be very clear from the graph that we could do much better if we introduced cubic and higher terms into the equation.

In Figs 8.10, 8.11, 8.12, 8.13 and 8.14 we have cubic quartic and quintic curves. The percentage fit offered by these equations are 98.5%, 98.8% and 99.0% respectively. The relatively small differences in percentage fit between these three

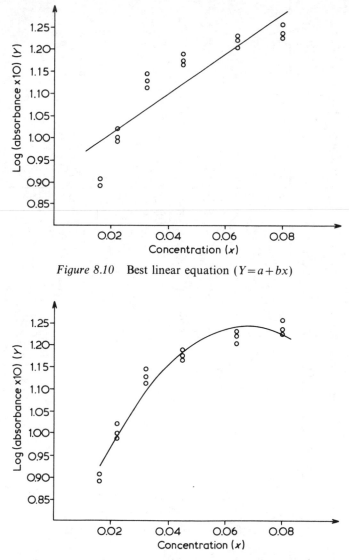

Figure 8.10   Best linear equation ($Y = a + bx$)

Figure 8.11   Best quadratic equation ($Y = a + bx + cx^2$)

equations is in contrast with the dramatic changes portrayed in the graphs. Having seen Fig. 8.14 who would dare to suggest that the quintic equation could be used for calibration purposes? But the unsuspecting amateur curve-fitter, who was using a polynomial program *without* such excellent graph-plotting facilities as those offered by this computer, would be unaware of the nonsense he was perpetrating. After all, the quintic equation does fit better than any other curve could. It passes through the mean *y* value at each of the six *x* values.

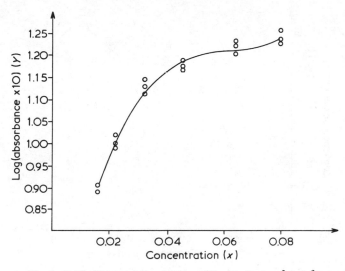

Figure 8.12   Best cubic equation ($Y = a + bx + cx^2 + dx^3$)

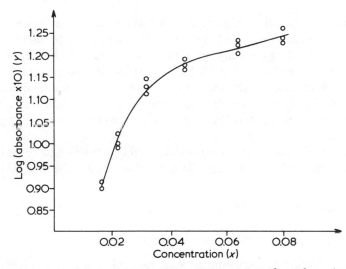

Figure 8.13   Best quartic equation ($Y = a + bx + cx^2 + dx^3 + ex^4$)

What conclusion are we to draw from this orgy of curve-fitting? If we are to use any one of the equations it would have to be the quartic. We could obtain predicted y values for a range of x values and use the confidence intervals to plot confidence bands onto Fig. 8.13. These could then be used in the reverse direction to get a confidence interval for the concentration of cuprammonium ion in an unknown sample.

Another alternative would be to put aside the Chief Chemist's obsession with

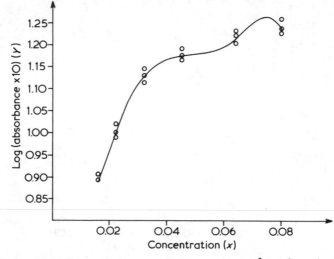

Figure 8.14   Best quintic equation ($Y = a + bx + cx^2 + dx^3 + ex^4 + fx^5$)

accommodating the whole range of concentrations into *one* calibration chart. He anticipates that concentrations between 0.002 M and 0.080 M might arise. What is wrong with dividing this range into three smaller ranges? Perhaps 0.002 M to 0.007 M, 0.007 M to 0.025 M and 0.025 M to 0.080 M would be suitable. Within each of these three intervals a linear equation or a quadratic equation might suffice. If the Chief Chemist were to adopt this policy, no one would applaud the sophistication of his approach, but Fig. 8.14 is a reminder of where sophistication can lead us.

If the Chief Chemist *must* have just one calibration chart to cover the whole range of concentration, perhaps he should consider the following simple procedure:

(a) Plot *one* point for each group of points in Fig. 8.9 using the mean absorbance for the group.
(b) Join the mean points together by straight lines.

Whilst this procedure is very simple, we can see in Fig. 8.15 that it compares well with the curve-fitting illustrated in Figs 8.10 to 8.14. Furthermore, it is much less dangerous.

8.6   Detecting curvature

We have emphasized the desirability of straight-line calibration and have pointed out the dangers of curve-fitting by means of high order polynomials. Even in those cases where the true relationship is obviously curved, we have suggested that it may be possible to fit a straight line that is satisfactory over a limited range

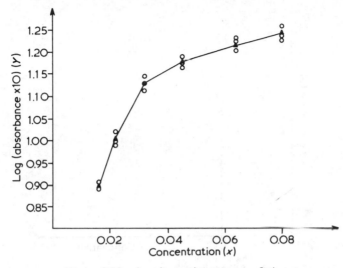

Figure 8.15   An alternative to curve-fitting

of concentration. In such cases the analyst must decide how wide a range of concentration he can use whilst still obtaining an adequate approximation to the curve by the straight line. Perhaps he would be wise to restrict his range so that a quadratic curve does not fit significantly better than a straight line. We will not discuss this point further but refer the reader to Davies and Goldsmith (1972) or Brownlee (1960).

Let us consider a rather different case in which the analyst is very confident that the relationship is linear at low concentrations, but he is equally sure that the straight line will bend at some higher and unknown concentration. In this situation the analyst might wish to determine the highest concentration for which the relationship is linear. This upper limit is represented by C in Fig. 8.16.

Several methods are available for detecting the upper limit of linearity. The one we will examine focuses attention on the *residuals* that are obtained when the best straight line is fitted to the calibration points. The reader may recall that we discussed residuals in Chapter 5 when we first used regression analysis. We calculated residuals for our calibration points using:

Residual = actual absorbance − predicted absorbance

For the calibration data in Fig. 8.1 we fitted the regression equation $y = 0.04 + 41.667x$ using ordinary regression in Chapter 7. (It is convenient to return to ordinary, rather than weighted, regression at this point in order to simplify the discussion.) Using the regression equation we can calculate the predicted absorbance for each point and hence the residual. This is done in Table 8.10.

Whenever we fit a regression line we can use the residuals to check on the

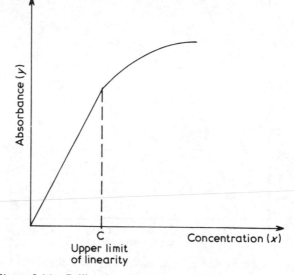

Figure 8.16 Calibration line curves at higher concentration

Table 8.10 Residuals from the calibration line

Concentration (x)	Actual absorbance (y)	Predicted absorbance (0.04 + 41.67x)	Residual
0.002	0.12	0.123	−0.003
0.003	0.14	0.165	−0.025
0.005	0.27	0.248	0.022
0.008	0.40	0.373	0.027
0.012	0.52	0.540	−0.020
			0.001

assumptions that underlie regression analysis. For example, underlying the calculation of a confidence interval for the true concentration of an unknown sample is the assumption that the random errors of the test method have a normal distribution. If this assumption is satisfied *and* we have fitted the right type of equation then we would expect the population of residuals to have a normal distribution. (We are regarding our five calibration samples as having been selected from a population of possible samples.) If we find, therefore, that the residuals do *not* appear to have come from a normal distribution we must suspect that either:

(a) the random errors have some other distribution, or
(b) the wrong type of equation has been fitted.

If we examine the residuals in Table 8.11 we find that some are negative and some are positive with the sums of the residuals equal to 0.001. This total would be equal to zero but for rounding errors. Plotting the residuals in a blob chart gives Fig. 8.17, which indicates a symmetrical distribution and could not be used to justify our rejecting the normal distribution.

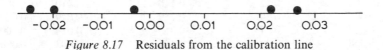

-0.02   -0.01   0.00   0.01   0.02   0.03

Figure 8.17   Residuals from the calibration line

A further assumption underlying the regression analysis is that the random errors should be independent of each other. If this assumption is satisfied we would not expect to find any serial pattern in the residuals. We would expect, therefore, that the positive and negative residuals would be mixed together in a random manner. If we found, for example, that a listing of residuals contained groups of + signs and groups of − signs we would suspect that either:

(a) errors in successive determinations were not independent, or
(b) some extraneous influence was causing a bias that persisted for limited periods, or
(c) we had fitted the wrong type of equation.

Concentrating on this last possibility we could account for the pattern of residuals (−, −, +, +, −) in Table 8.10 as indicating a need for a different equation. Perhaps this data needs a curve over the whole range of concentration or perhaps the point with highest concentration needs to be eliminated. Figure 8.2 offers support for *either* of these possibilities. A third possibility is that the pattern in the residuals is simply due to chance. We have only five points and with so little data it is difficult *not* to see a pattern when adopting such a subjective approach.

Let us make a more objective exploration of the possibility that 'the true relationship is linear for low concentrations but the highest concentration used (0.012 M) is beyond the upper limit of linearity'. We will fit a straight line to the first four points *only*, and then we will examine the residuals from all five samples. If the residual from the fifth point, with highest concentration, does not appear to share a normal distribution with the other four residuals we will conclude that 0.012 M is beyond the upper limit. The fitted equation is $y = 0.0129 + 48.810x$, and the residuals are listed in Table 8.11.

Because we have used only four points to fit the calibration line we must not expect the five residuals to have a total of zero. The first four residuals will sum to zero (neglecting the small deviation due to rounding errors) but we would expect the fifth residual to be *negative* if a concentration of 0.012 M is beyond the limit of linearity. As we can see in Fig. 8.18 the fifth residual is very negative and gives a point far removed from the other four.

Statistics for Analytical Chemists

Table 8.11   Residuals from a four-point calibration

Concentration (x)	Actual absorbance (y)	Predicted absorbance (0.0129 + 48.810x)	Residual	Squared residual
0.002	0.12	0.110	0.010	0.000 100
0.003	0.14	0.159	−0.019	0.000 361
0.005	0.27	0.257	0.013	0.000 169
0.008	0.40	0.403	−0.003	0.000 009
0.012*	0.52*	0.599	−0.079	
				0.000 639

* Not used when fitting the calibration line.

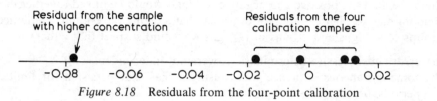

Figure 8.18   Residuals from the four-point calibration

Figure 8.18 appears to give a very strong indication that the residual from the fifth sample did not come from the same normal distribution as the other four residuals. Before we conclude that a concentration of 0.012 M lies above the limit of linearity, we would be wise to carry out an appropriate significance test. 'Obvious patterns' can easily arise by chance with so little data. We could use Dixon's test to see if the fifth residual could reasonably be branded as an outlier. Dixon's test is intended to be used in situations where we identify the most extreme values *after* the data has been gathered. In this situation we know *in advance* which residual we are interested in.

We will consider an alternative test. This is an adaptation of the widely used 'two-sample *t*-test' which is similar to the one-sample *t*-test that we used in Chapter 4. We will use μ_A to represent the mean of residuals for all samples with concentration *above* the limit of linearity and μ_B for the mean of residuals *below* the limit.

Null hypothesis – $\mu_A = 0$

Alternative hypothesis – $\mu_A < 0$

$$\text{Test statistic} = \frac{\text{residual from the sample with high concentration}}{(\text{residual standard deviation}) \sqrt{(n+1)/n}}$$

$$= \frac{0.079}{0.0179\sqrt{(5/4)}}$$

$$= 3.95$$

Critical values – from the *t*-table with 2 degrees of freedom for a one-sided test:

> 2.92 at the 5% significance level
> 6.97 at the 1% significance level

Decision – We reject the null hypothesis at the 5% level.

Conclusion – We conclude that the linear relationship that applies between 0.002 M and 0.008 M does *not* extend to a concentration of 0.012 M.

It is appropriate to carry out a *one-sided* test becase we expect that, when we reach a concentration at which the linear relationship ceases to apply, the new relationship will have a *decreased* slope. We expect a change of slope in *one* direction.

Having concluded that a concentration of 0.012 M is beyond the upper limit of linearity, we could now focus attention on the sample which has a concentration of 0.008 M. After fitting a line to the three lowest points (0.002 M, 0.003 M, 0.005 M) we could examine the residual from the fourth point (0.008 M). In the *t*-test we would be unable to reject the null hypothesis and would conclude that the linear region did include 0.008 M.

When using any significance test successively, as we have done in this situation, we should bear in mind that the significance level may not be what it first appears. If, for example, we use a 5% significance level in each of a succession of *t*-tests then the overall significance level for the whole series of tests will be greater than 5%. It is advisable, therefore, to use a 1% significance level when following this procedure for detecting curvature. Furthermore, the analyst would be wise to use a larger number of samples than we have used in this example and to adopt roughly equal spacing in concentration.

8.7 Summary

In this chapter we have considered various complications that can prevent the use of simple linear regression for fitting a calibration line. We explored the use of weighted regression where the precision of the test method depends upon the concentration. We saw that this gave very different confidence intervals to those given by ordinary regression. We briefly examined some of the dangers inherent in the use of curve-fitting programs and advised the use of linear regression wherever possible. Finally we looked at a simple method for detecting the onset of curvature in cases where the relationship is known to be linear at low concentrations.

In the next chapter we will return to an earlier theme when we consider the estimation of the reproducibility of a test method. Many interesting problems arise when we attempt to separate the variability within laboratories from the variability between laboratories. A very powerful statistical technique will be introduced to cope with these difficulties.

Problems

(1) Prior to the introduction of a new test method an investigation was carried out to explore the error structure of the method. The researcher concluded that the precision of the test method was not constant and, for concentrations in the range 10 to 100 p.p.m., the following relationship was established:

$$\left(\begin{array}{c}\text{SD of repeat determinations}\\ \text{(absorbance units)}\end{array}\right) = 0.997 + 0.0206\left(\begin{array}{c}\text{concentration}\\ \text{(p.p.m.)}\end{array}\right)$$

At a later date the absorbances of four standard samples were measured in order to calibrate the test method. The results are:

Concentration (p.p.m.)	10	40	70	100
Absorbance	5	17	35	45

(a) Use the equation to calculate a standard deviation for each level of concentration used in the calibration.
(b) Use the standard deviations from (a) to calculate a weight for each of the four points.
(c) Calculate the slope and intercept of the weighted least squares regression line for the calibration data.
(d) Calculate the residual standard deviation.
(e) Calculate a confidence interval for the true concentration of a sample that gave an absorbance measurement of:

   (i) 5
   (ii) 45

(2) An analyst is confident that the calibration line for a particular test method will be linear at low concentrations. In order to find the upper limit of the linear range he measures the absorbance of ten samples containing increasing concentration of a particular substance. The results are:

Concentration in p.p.m. (x)	2	4	6	8	10	12	14	16	18	20
Absorbance (y)	14	21	25	33	36	42	50	54	57	61

Using a pocket calculator the analyst fits a regression equation by ordinary least squares to the first nine points. Then he fits an equation to the first eight points, then the first seven points, etc. The equations are:

Range of concentration (p.p.m.)	Equation	RSD
2 to 18 inclusive	$y = 9.389 + 2.750x$	1.409
2 to 16 inclusive	$y = 8.714 + 2.851x$	1.166
2 to 14 inclusive, etc.	$y = 8.571 + 2.875x$, etc.	1.262

Carry out a succession of t-tests in order to establish the upper limit of the linear range.

————9————
Reproducibility (1)

9.1   Introduction

In Chapter 3 we defined *repeatability* in terms of the difference one is likely to find between two determinations made in the *same* laboratory. It was assumed that the two determinations were made by the same operator using the same apparatus and that the work was carried out over a short period of time. Because of random errors, the difference between two determinations is unlikely to be zero but there is a 95% chance that it will be less than the repeatability.

If we take two determinations which were made in *different* laboratories (which would imply different operators and apparatus), we would not be surprised to find a *larger* difference between them. In this chapter we will examine the concept of *reproducibility* which gives us a 95% upper limit for the difference between two determinations produced in *different* laboratories. Clearly this idea is very important whenever a standardized test method is used throughout the many laboratories of a large company or within the many companies of a particular industry. It is *so* important that British Standard 5497: *Precision of Test Methods* advises us on the conduct of *precision experiments* which are designed to produce estimates of the repeatability and the reproducibility of a standardized procedure.

In this chapter we will introduce the results of a precision experiment. The analysis of this rather large set of data will be spread over several chapters but we will first bring to bear upon it a very powerful statistical technique known as *analysis of variance*. The use of this technique will enable us to separate the variability *within* laboratories from the variability *between* laboratories. This is a prerequisite to calculating the reproducibility of the standardized test method.

9.2   A precision experiment

The Chief Chemist of Indichem Ltd wishes to carry out a full-scale investigation to assess the repeatability and the reproducibility of the standard method for the determination of ammonia content of ammonia solution as per BS 4651. This British Standard is in use in many companies and authorities throughout the UK

126

and is also used in other countries. The Chief Chemist is only concerned, however, with its use within the thirty laboratories of Indichem and he designs the following experiment which he hopes will allow him to estimate the repeatability and reproducibility of the standard method.

Since ammonia solution is used in many different concentrations throughout the company he selects four levels for consideration and he orders the preparation of 2 litres of ammonia solution at concentrations of 10%, 15%, 20% and 25% by mass. From the thirty laboratories, four are selected at random and each of the four laboratories is sent five samples at each of the four concentrations. Each of the four laboratories prepares its own reagents and the twenty determinations from each laboratory are listed in Table 9.1.

Table 9.1   Results of the precision experiment

Level	A		B		C		D	
1 (10%)	10.0	10.0	9.9	10.0	10.1	10.3	9.9	10.0
	10.1	9.9	10.0	10.1	10.0	10.0	9.8	9.9
	10.0		10.0		10.1		9.9	
2 (15%)	15.2	14.9	15.0	15.3	14.9	15.1	14.8	15.0
	14.8	15.1	15.3	14.9	15.1	15.2	14.7	14.8
	15.0		15.0		15.2		14.7	
3 (20%)	19.7	20.1	20.2	20.3	20.3	20.0	19.9	19.8
	19.7	19.7	20.2	20.0	19.9	20.3	19.9	19.9
	19.8		20.3		20.0		20.0	
4 (25%)	24.8	25.1	24.7	25.1	25.6	25.3	25.4	25.0
	25.2	24.8	24.5	24.9	25.1	25.6	25.4	25.2
	25.1		24.8		25.4		25.0	

In a later chapter we will estimate the repeatability and the reproducibility of the test method at each of the four levels, separately. We will also investigate the possibility that repeatability and/or reproducibility may vary from level to level in a systematic manner. In *this* chapter we will confine our attention to the third level only (20% concentration).

Using the twenty determinations made by the four laboratories on the samples at 20% concentration we will attempt to quantify the variability *within* the four laboratories and the variability *between* them.

9.3   Variability between and within laboratories

The data in Table 9.2 has been extracted from the third row of Table 9.1. To simplify the calculations we are about to perform, we will reduce the size of the numbers by subtracting a constant from each. Since the smallest of the determinations is 19.7, this is a suitable number to subtract. Note that carrying out this subtraction does not change the *variability* of the data.

Table 9.2   Results of the precision experiment – level 3 only

	Laboratory			
	A	*B*	*C*	*D*
Determinations minus 19.7	0.0 0.4	0.5 0.6	0.6 0.3	0.2 0.1
	0.0 0.0	0.5 0.3	0.2 0.6	0.2 0.2
	0.1	0.6	0.3	0.3
Mean	0.10	0.50	0.40	0.20
Standard deviation	0.173	0.122	0.187	0.071

Using the data in Table 9.2 we can estimate:

(a) The true concentration of ammonia (μ) as determined by this test method.
(b) The within-laboratory variance (σ_w^2).
(c) The between-laboratories variance (σ_b^2).

If the samples have been correctly prepared and if the test method were not biased then μ would be equal to 20.0. To estimate μ we will use the mean of the twenty determinations. Should this prove to be significantly different from 20.0 there would be cause for concern, but we could still estimate the repeatability and reproducibility of the test method.

The within-laboratory variance (σ_w^2) is a measure of the variability we would get in the determinations if they were all carried out in the *same* laboratory. We will assume that determinations made in the same laboratory were made by the same operator using the same equipment. The variability encompassed by σ_w^2 results from the random error that is present in all test measurements. Clearly the within-laboratory variance is related to the repeatability of the test method, and the square root of the within-laboratory variance would be identical to the error standard deviation (ESD) that we have used in earlier chapters.

The between-laboratories variance (σ_b^2) is a measure of the *extra* variability that is introduced if determinations are carried out in different laboratories. This extra variability will include the systematic errors which cause a laboratory to be biased and additional random errors which would not be present under repeatability conditions.

Before we get down to the calculations which produce estimates of μ, σ_w^2, and σ_b^2 let us examine an alternative explanation which might help to clarify just what it is that we are estimating. Statisticians often use models to represent complex situations. Though these models are rather abstract, some scientists find them very helpful especially when discussing the *assumptions* which underlie a statistical technique. A suitable model for the situation which gave rise to the data in Table 9.2 would be:

$$y = \mu + \beta + \delta$$

in which y = a determination of ammonia content

   μ = the true mean ammonia content of the samples

   β = a deviation from the true mean because of systematic error peculiar to the *laboratory*

   δ = a further deviation due to the random error associated with the test method.

There is, of course, only one value for μ. There are four values for β; one for each laboratory. There are twenty values for δ; one for each determination. If we knew the twenty values of δ, the four values of β and the value of μ we could substitute them into the model to get the twenty determinations. Obviously one can never do this in practice. The variances that we hope to estimate can also be related to the model, as follows:

σ_b^2 = the variance of the β values for all thirty laboratories in the population;
σ_w^2 = the variance of the δ values for all possible random testing errors.

The graphs which follow may also help you to understand the nature of the within-laboratory variance and the between-laboratories variance. In Fig. 9.1 is plotted the data of Table 9.2. We can see that, within each laboratory the determinations are scattered about the mean for that laboratory. We can also see that the four laboratory means (0.1, 0.5, 0.4 and 0.2) are scattered about the overall mean (0.3). In Fig. 9.1 we have variation *within* each laboratory and we have variation *between* laboratories. The presence of *both* sources of variation makes it difficult to appreciate the two variances that we wish to estimate.

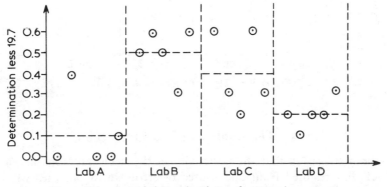

Figure 9.1 Actual determinations of ammonia content

Figures 9.2 and 9.3 represent hypothetical situations in which one of the two variances is equal to zero. In Fig. 9.2 the within-laboratory variance is equal to zero so that all the determinations from any particular laboratory are equal to each other, but the laboratory means are equal to those in Fig. 9.1. In Fig. 9.3 the between-laboratories variance is equal to zero and we find no difference between the laboratory means, but the within-laboratory spreads are the same as those in Fig. 9.1. In both Fig. 9.2 and Fig. 9.3 there is *less* variability than we have in the real data of Fig. 9.1.

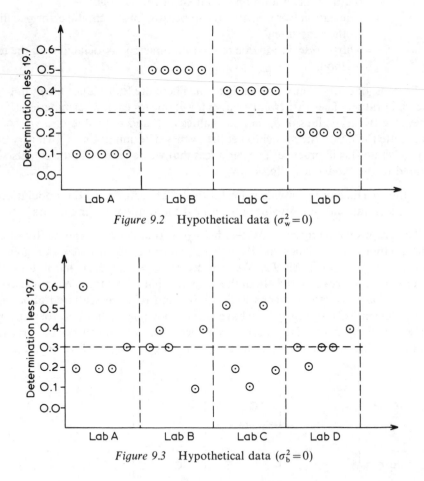

Figure 9.2   Hypothetical data ($\sigma_w^2 = 0$)

Figure 9.3   Hypothetical data ($\sigma_b^2 = 0$)

9.4   The analysis of variance table

It has been suggested that the variability in the determinations in Table 9.2 (Fig. 9.1) has resulted from one source of error which operates within any laboratory and a second source of error which causes differences between laboratories. We will now measure the *total* variability in Table 9.2 and break it down into two components. These two components can then be used to estimate σ_w^2 and σ_b^2.

In Chapter 2 we introduced the variance and the standard deviation as measures of spread or scatter. In subsequent chapters we made frequent use of the standard deviation as an indication of precision, but we made less use of the variance. We noted that standard deviations must never be added and for this reason we will be forced to speak of variances when discussing reproducibility. Before this, however, we must retreat even further and make use of the sum of squares if we are to separate the within-laboratory variation and the between-laboratories variation. (In Chapter 2, you may recall, we defined the variance as being equal to the sum of squares divided by its degrees of freedom.) We will calculate:

(a) the total sum of squares;
(b) the within-laboratory sum of squares;
(c) the between-laboratories sum of squares.

The calculations are set out in tabular form in Table 9.3, which contains the data from Table 9.2 and makes use of the following formulae:

$$\text{Total sum of squares} = \sum(\text{determination} - \text{overall mean})^2$$
$$\text{Within-lab sum of squares} = \sum(\text{determination} - \text{lab mean})^2$$
$$\text{Between-labs sum of squares} = \sum[\text{no. of determinations in lab}) \times (\text{lab mean} - \text{overall mean})^2]$$

Concerning the sums of squares the following points are worthy of note:

(a) The total sum of squares (0.84) is equal to the within-laboratory sum of squares (0.34) *plus* the between-laboratories sum of squares (0.50). This is not a mere coincidence.
(b) The within-laboratory sum of squares would have a large value if the determinations were widely scattered about the laboratory means.
(c) The within-laboratory sum of squares would be equal to zero for the hypothetical data in Fig. 9.2.
(d) For any set of data which gives a within-laboratory sum of squares equal to zero we can conclude without reservation that the within-laboratory variance (σ_w^2) is equal to zero. (Since this is such an unlikely occurrence we might suspect that the data had been manipulated.)
(e) The between-laboratories sum of squares would have a large value if the laboratory means were widely scattered about the overall mean.
(f) The between-laboratories sum of squares would be equal to zero if the four laboratory means were equal to each other; this would be so with the hypothetical data in Fig. 9.3 but would be extremely unlikely to occur in practice.

Table 9.3 Calculation of sums of squares

1	2	3	4	5	6	7	8	9	10
Laboratory	Determination	Determination minus overall mean	Column 3 squared	Laboratory mean	Determination minus lab mean	Column 6 squared	Determination replaced by lab mean	Lab mean minus overall mean	Column 9 squared
A	0.0	-0.3	0.09		-0.1	0.01	0.1	-0.2	0.04
	0.4	0.1	0.01		0.3	0.09	0.1	-0.2	0.04
	0.0	-0.3	0.09	0.1	-0.1	0.01	0.1	-0.2	0.04
	0.0	-0.3	0.09		-0.1	0.01	0.1	-0.2	0.04
	0.1	-0.2	0.04		0.0	0.00	0.1	-0.2	0.04
B	0.5	0.2	0.04		0.0	0.00	0.5	0.2	0.04
	0.6	0.3	0.09		0.1	0.01	0.5	0.2	0.04
	0.5	0.2	0.04	0.5	0.0	0.00	0.5	0.2	0.04
	0.3	0.0	0.00		-0.2	0.04	0.5	0.2	0.04
	0.6	0.3	0.09		0.1	0.01	0.5	0.2	0.04
C	0.6	0.3	0.09		0.2	0.04	0.4	0.1	0.01
	0.3	0.0	0.00		-0.1	0.01	0.4	0.1	0.01
	0.2	-0.1	0.01	0.4	-0.2	0.04	0.4	0.1	0.01
	0.6	0.3	0.09		0.2	0.04	0.4	0.1	0.01
	0.3	0.0	0.00		-0.1	0.01	0.4	0.1	0.01
D	0.2	-0.1	0.01		0.0	0.00	0.2	-0.1	0.01
	0.1	-0.2	0.04		-0.1	0.01	0.2	-0.1	0.01
	0.2	-0.1	0.01	0.2	0.0	0.00	0.2	-0.1	0.01
	0.2	-0.1	0.01		0.0	0.00	0.2	-0.1	0.01
	0.3	0.0	0.00		0.1	0.01	0.2	-0.1	0.01
Total	6.0	0.0	0.84		0.0	0.34	6.0	0.0	0.50
	Mean = 0.3		Total SS			Within labs SS			Between labs SS

(g) Iɪ. practice the laboratory means and the overall mean would not be convenient numbers and the following formulae would be more useful:

Total sum of squares = (SD of all determinations)^2 × (degrees of freedom)

Within-lab sum of squares = $\sum$[(SD of determinations in one lab)^2 × (degrees of freedom)]

Between-labs sum of squares = (SD of lab means)^2 × (degrees of freedom) × (no. of determinations per lab)

In each formula the degrees of freedom are obtained by subtracting one from the number of numbers that went into the standard deviation.

Having calculated the sums of squares it is conventional to draw up an analysis-of-variance table like Table 9.4.

Table 9.4   Analysis of variance table

Source of variation	Sum of squares	Degrees of freedom	Mean square
Between laboratories	0.50	3	0.166 7
Within laboratory	0.34	16	0.021 25
Total	0.84	19	

Within the analysis-of-variance table we divide a sum-of-squares by its degrees of freedom to obtain a mean square. Clearly a mean square is similar to a variance and we will use the mean squares in Table 9.4 to estimate the two variances σ_w^2 and σ_b^2. The degrees of freedom in Table 9.4 are obtained by applying the following rules:

(a) The total degrees of freedom (19) is one less than the total number of determinations (20).
(b) The between-laboratories degrees of freedom (3) is one less than the number of laboratories (4).
(c) Within each laboratory we have five determinations. Subtracting one from this figure gives us four degrees of freedom within *each* laboratory. Since there are four laboratories and four degrees of freedom within each, the 'within-laboratory degrees of freedom' is equal to four multiplied by four which is sixteen.

The degrees of freedom in Table 9.4 can also be calculated by counting the number of *independent* entries in certain columns of Table 9.3. Column 3 contains only nineteen independent entries since the twenty entries must sum to zero. Column 6 contains only sixteen independent entries since the sum for each laboratory must be zero. Column 9 contains only three independent entries since there are four different entries which must sum to zero.

Having completed the analysis of variance table we are now in a position to estimate the within-laboratory variance σ_w^2 and the between laboratory variance σ_b^2.

9.5 Variance estimates and their significance

Consider the within-laboratory mean square in Table 9.4. The data we have analysed has given us a value of 0.021 25 for this mean square. A different set of data would have given us a different value but the lowest figure we *could* have obtained for the within-laboratory mean square is zero. This could only arise if there was no random error associated with the test method for determination of the ammonia content, i.e. if each value of δ in the model was equal to zero which would imply that $\sigma_w^2 = 0$. (An alternative explanation is that there *is* random error associated with the procedure but, by chance, each of our twenty determinations has a random error of zero – a very remote possibility.)

The within-laboratory mean square in Table 9.4 has resulted *solely* from the variability within the determination process. The bias of one or more laboratories cannot *possibly* have contributed to the calculated value of 0.021 25. We therefore estimate that $\sigma_w^2 = 0.021\ 25$.

> Since the variation in the five determinations from any one laboratory is entirely due to testing error, the *within-samples mean square* is used as an estimate of the within laboratory variance.

The within-laboratory variance (σ_w^2) could also be estimated by combining the standard deviations in Table 9.2. To perform the calculation we could adapt the formula introduced in Chapter 6:

$$\text{Combined standard deviation} = \sqrt{\left\{ \frac{\sum[(\text{d.f.})\,(\text{SD})^2]}{\sum(\text{d.f.})} \right\}}$$

As we are estimating a variance the square root will be removed to give:

$$\text{Estimate of within-laboratory variance} = \frac{\sum[(\text{d.f.})\,(\text{SD})^2]}{\sum(\text{d.f.})}$$

$$= \frac{4(0.173)^2 + 4(0.122)^2 + 4(0.187)^2 + 4(0.071)^2}{4+4+4+4}$$

$$= 0.021\ 21$$

This estimate would be in perfect agreement with the earlier estimate (0.021 25) but for rounding errors in the standard deviations in Table 9.2. Beware of premature rounding.

When using data from a sample to estimate one or more characteristics of a population, there is always a danger that we will produce an estimate of something which does not exist. It is possible, for example, that either $\sigma_w^2 = 0$ or $\sigma_b^2 = 0$ or both.

9.5.1   *Consider the hypothesis* $\sigma_w^2 = 0$

We can reject this without hestitation since, if it were true, all the determinations from a particular laboratory would have the same value and our data would resemble the hypothetical data in Fig. 9.2.

9.5.2   *Consider the hypothesis* $\sigma_b^2 = 0$

If this were true we would have no long term differences between the thirty laboratories from which the four laboratories were sampled. Nonetheless, due to random errors in the determinations (since $\sigma_w^2 \neq 0$), we might find differences between laboratories in the short term but these differences would be relatively small. (Speaking more generally, you would not expect the group means to be *exactly* equal if you made twenty determinations on the same sample and put them into four groups of five.) Even if the between laboratory variance (σ_b^2) were equal to zero we would not therefore expect the four laboratory means to be exactly equal. (In practice, the hypothetical data portrayed in Fig. 9.3 would be very unlikely to arise.) As we actually *found* differences between our laboratory means (0.10, 0.50, 0.40 and 0.20) we must ask ourselves 'Is the variability between laboratories so great that we can reject the hypothesis $\sigma_b^2 = 0$?'

Looking at this problem in terms of the model, the hypothesis ($\sigma_b^2 = 0$) would imply that all values of β were equal to zero and all the variability in the data would be due to differences in the twenty values of δ in our data. When calculating the laboratory means we put these values of δ together in groups of five and therefore we would expect to get laboratory means that were not equal, even if $\sigma_b^2 = 0$.

To test the hypothesis '$\sigma_b^2 = 0$' we use an *F*-test as follows:

Null hypothesis – There is no long term variation between laboratories ($\sigma_b^2 = 0$).

Alternative hypothesis – There is long term variation between laboratories ($\sigma_b^2 > 0$).

$$\text{Test statistic} = \frac{\text{between-labs mean square}}{\text{within-lab mean square}}$$

$$= \frac{0.166\,7}{0.021\,25} = 7.84$$

Critical values – from the one-sided *F*-table with 3 and 16 degrees of freedom are:

> 3.25 at the 5% significance level
> 5.32 at the 1% significance level

Decision – We reject the null hypothesis at the 1% significance level.

Conclusion – We conclude that there are long-term differences between laboratories.

Two points should be noted concerning the above significance test:

(a) If the null hypothesis were true we would expect the between-laboratories mean square to be approximately equal to the within-laboratory mean square which would result in the value of the test statistic being approximately equal to one. If the test statistic is significantly greater than one therefore, we reject the null hypothesis.

(b) Critical values for this decision are taken from Table C using 3 and 16 degrees of freedom for a *one-sided test*. (We *always* use a one-sided *F*-test when comparing two mean squares from an analysis-of-variance table.)

Now that we are confident that the between-laboratories variance is *not* equal to zero, how are we to estimate its value? We cannot simply use the between-laboratories mean square as an estimate since *both* the between-laboratories variability *and* the within-laboratory variability have contributed to this mean square.

We estimate the between-laboratories variance σ_b^2 as (between-laboratories mean square – within-laboratory mean square)/(number of determinations per laboratory)

$$\text{estimate of } \sigma_b^2 = (0.1667 - 0.021\,25)/5$$
$$= 0.029$$

Note: If the between-laboratories mean square had not been greater than the within-laboratory mean square we would *not* have carried out the *F*-test; we would have accepted that σ_b^2 could well be equal to zero. Alternatively, if we had carried out the *F*-test and failed to reject the null hypothesis we would again conclude that σ_b^2 might be equal to zero. In either case we would then use the total mean square as an estimate of the *within*-laboratory variance σ_w^2.

Having estimated the two variances we are now in a position to calculate a confidence interval for the true ammonia content of the samples. The best estimate we have of the true ammonia content is the overall mean of the twenty determinations which is 20.0. Had we selected different laboratories for inclusion in the precision experiment or carried out the experiment at a different time we would almost certainly have obtained a different value for the overall mean. The width of the confidence interval for the true mean will depend upon the variability that one might expect in the overall mean determination if the experiment were repeated many many times.

A confidence interval for the true mean determination is given by:

$$\text{Overall mean} \pm t \sqrt{\left(\frac{\text{estimate of } \sigma_b^2}{\text{no. of labs}} + \frac{\text{estimate of } \sigma_w^2}{\text{total no. of determinations}} \right)}$$

where t is taken from the two-sided t-table with appropriate degrees of freedom

Our estimate of the between-laboratories variance (σ_b^2) has 3 degrees of freedom and our estimate of the within-laboratory variance (σ_w^2) has 16 degrees of freedom. The degrees of freedom needed to obtain the t value in the above formula, will lie between these two figures (3 and 16). A formula for calculating the degrees of freedom in this situation is given in Wilson (1979). When the between-laboratories mean square (BLMS) is much greater than the within-laboratory mean square (WLMS) a close approximation is given by:

Degrees of freedom

$$= (\text{no. of labs} - 1)\left[1 + (\text{no. of determinations per lab} - 1)\frac{\text{WLMS}}{\text{BLMS}} \right]^2$$

$$= (4 - 1)\left(1 + (4)\frac{0.021\,25}{0.166\,7} \right)^2$$

$$= 6.84$$

Using 7 degrees of freedom and a 5% significance level we get a two-sided t value of 2.36, giving the 95% confidence interval:

$$20.0 \pm 2.36\sqrt{(0.029/4 + 0.021\,25/20)}$$
$$= 20.0 \pm 2.26\sqrt{(0.007\,25 + 0.001\,062\,5)}$$
$$= 20.0 \pm 0.215$$

So far we have regarded the four laboratories as being a random sample drawn from the thirty laboratories within Indichem Ltd. We will continue to regard them in this light when we later discuss the reproducibility of the test method. On the other hand we could examine each laboratory in its own right and ask 'What would be the mean determination for laboratory X if it had produced a very large number of determinations, rather than just five?'

> A confidence interval for the true mean determination of a particular laboratory is given by:
>
> laboratory mean $\pm t\sqrt{[}$ estimate of $\sigma_w^2/$(no. of determinations in that lab)$]$
>
> where t is taken from the two-sided t-table using the within-laboratory degrees of freedom.

Substituting into the above formula $t_{16} = 2.12$, and using the estimate of σ_w^2 (i.e. 0.021 25) we get the confidence intervals in the penultimate column of Table 9.5.

Table 9.5   Confidence intervals of laboratory means and SDs

Laboratory	Sample Mean	SD	Confidence intervals for population Mean	SD		
A	$\bar{x}_A = 19.8$	$s_A = 0.173$	$\mu_A = 19.8 \pm 0.14$	$\sigma_A = 0.108$	to	0.223
B	$\bar{x}_B = 20.2$	$s_B = 0.122$	$\mu_B = 20.2 \pm 0.14$	$\sigma_B = 0.108$	to	0.223
C	$\bar{x}_C = 20.1$	$s_C = 0.187$	$\mu_C = 20.1 \pm 0.14$	$\sigma_C = 0.108$	to	0.223
D	$\bar{x}_D = 19.9$	$s_D = 0.017$	$\mu_D = 19.9 \pm 0.14$	$\sigma_D = 0.108$	to	0.223

The confidence intervals for the laboratory standard deviations in Table 9.5 are the same for each laboratory. We are assuming that the four laboratories would be equally variable in the long term, but more will be said about this later when we discuss the assumptions underlying analysis of variance. Each confidence interval in the final column is based on the estimate of $\sigma_w (\sqrt{0.021\ 25} = 0.146)$ and multipliers from Table F.

The confidence intervals in Table 9.5 are rather wide. You may be surprised that the use of a sophisticated statistical technique such as analysis of variance should produce such disappointingly wide confidence intervals. We must not forget, however, that the estimates of means and variances are based on very little information. Only four laboratories were used in the experiment and only five determinations were made in each laboratory. If we had carried out a larger experiment we would probably have obtained narrower intervals. Using the two variance estimates from the completed experiment we can calculate the width of a confidence interval that we could expect to get from experiments of different size. The formula for an approximate confidence interval for the true determination, that we used earlier, gives the intervals in Table 9.6 when we substitute different numbers of laboratories and different numbers of determinations.

If we intended to carry out an experiment with the *sole* intention of estimating the true ammonia content of a sample, then we can see from Table 9.6 that it

Table 9.6   Width of 95% confidence interval for true ammonia content

Number of determinations in each laboratory	Number of laboratories									
	1		2		4		8		16	
1	0.89	1.05	0.63	0.74	0.45	0.53	0.32	0.38	0.22	0.26
2	0.80	0.94	0.57	0.67	0.40	0.47	0.28	0.33	0.20	0.24
4	0.74	0.87	0.52	0.61		0.44	0.26	0.31	0.18	0.21
8	0.72	0.85	0.51	0.60	0.36	0.42	0.25	0.30	0.18	0.21
16	0.69	0.81	0.49	0.58	0.35	0.41	0.24	0.28	0.17	0.20

would be 'better' to carry out two determinations in each of four laboratories than to carry out four determinations in each of two laboratories, for example. (Better in the sense that the first alternative could be expected to yield a confidence interval (± 0.20) which was narrower than one from the second alternative (± 0.26).)

It is perhaps easier to appreciate the relative benefits of increasing the number of laboratories (p) in the experiment or increasing the number of determinations made in each laboratory (n) if the information in Table 9.6 is presented in graphical form. Figure 9.4 facilitates the comparison of different experimental designs, but any realistic comparisons would need to take into account the costs of sampling and testing.

The precision experiment that was introduced in this chapter was *not* carried out with the sole purpose of estimating the true ammonia content of a sample of

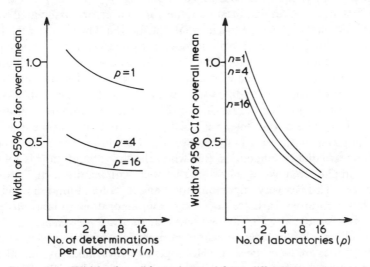

Figure 9.4   Width of confidence interval from different experiments

ammonia solution. The main purpose, you may recall, was to estimate the repeatability and the reproducibility of the test method. We will return to this main theme after we have examined certain assumptions which underlie the analysis of variance technique that we have just used.

9.6 Assumptions underlying analysis of variance

What we have referred to in this chapter as 'the analysis of variance technique' is just *one* of many methods of analysis which come under the umbrella of 'analysis of variance'. In the next chapter we will examine other situations in which different versions of analysis of variance are appropriate. Some of these situations are more complex and the power of analysis of variance will be more apparent. As the spread of computing facilities makes the more powerful statistical techniques widely available, it is unfortunately true that these techniques are being used by people who are unaware of the assumptions on which they are based. We will now examine the assumptions which underlie the analysis of variance that we have already carried out.

With reference to our model $x = \mu + \beta + \delta$, the assumptions can be summarized most succinctly as follows:

(a) It is assumed that each value of δ is taken at random from a normal distribution which has a mean equal to zero and a variance equal to σ_w^2.
(b) It is assumed that each value of β is taken at random from a normal distribution which has a mean equal to zero and a variance equal to σ_b^2.

These assumptions are expressed in the language of the statistician. Just what do they mean in practical terms, and are they likely to be satisfied by the precision experiment that yielded the data in Table 9.2?

The first assumption requires that the random errors associated with the standard test method give rise to a normal distribution. The harsh truth about the normal distribution is that it is a model devised by mathematicians and it only fits *exactly* to certain abstract situations. Nonetheless the normal distribution has been found to be an adequate model of many practical situations in which numerous small errors are adding together and it might be reasonable to assume a normal distribution for the random errors in the determination of ammonia concentration. The phrase 'mean equal to zero' raises no problem as we can include any systematic error in the values of β. The phrase 'variance equal to σ_w^2' might appear equally innocuous, but when the first assumption is taken as a whole it implies that we would find the *same variance* in each of the thirty laboratories. This is a very important point and it is often found in practice that the within-laboratory variance does vary from laboratory to laboratory. If the laboratories are equally variable we say that they are *homogeneous*. If, on the other hand, some laboratories appear to be more variable than others we say that the laboratories are *heterogeneous*. 'Homogeneity of variance' as it is called, is so important that significance tests have been devised which allow us to check whether or not it is safe to assume that the same value of within-laboratory variance would be found in all laboratories from which our sample of four laboratories was taken.

Three such tests are in common use. They are:

(a) Hartley's variance ratio test.
(b) Bartlett's variance homogeneity test.
(c) Cochran's maximum variance test.

It is the last of these three which is recommended by BS 5497 for use in the analysis of precision experiments. As you will recall from Chapter 6 the calculation of the test statistic for Cochran's test involves dividing the *largest* of the laboratory variances by the *sum* of the laboratory variances.

Null hypothesis – All the laboratories have the same value of within-laboratory
         variance.

Alternative hypothesis – One laboratory is more variable than the others.

$$\text{Test statistic} = \frac{\text{largest variance}}{\text{sum of the variances}}$$

$$= \frac{(0.187)^2}{(0.173)^2 + (0.122)^2 + (0.187)^2 + (0.071)^2}$$

$= 0.41$ (Using the standard deviations from Table 9.2.)

Critical values – from Table H for four laboratories with five determinations in
         each, are:

         0.629 at the 5% significance level
         0.721 at the 1% significance level

Decision – We cannot reject the null hypothesis.

Conclusion – We conclude that the 'homogeneity of variance' assumption has not
         been violated.

Having carried out the homogeneity of variance test let us examine the second of the two assumptions which underlie the simple analysis of variance. The phrase 'is taken at random' implies that the four laboratories chosen for inclusion in the experiment should have been selected at random from the thirty laboratories within the company. The four laboratories should constitute a random sample from the population of thirty laboratories. (Strictly speaking this population should be infinite but in practice it never will be.) The phrase 'mean equal to zero' implies that the bias of individual laboratories should average out over the whole company with some laboratories showing a negative bias and others a positive bias. If this condition were *not* satisfied it would imply the presence of a bias in the standardized method of determination. Ideally such a possibility would have been eliminated before carrying out a precision experiment but the presence of such a company-wide bias would not necessarily invalidate our variance estimates.

This discussion of the assumptions underlying analysis of variance has, of necessity, been somewhat abstract. It has been related to the model, which you might have already rejected as being somewhat remote from the reality of the laboratory. To summarize the discussion on a more practical note, let it be emphasized that the two assumptions most likely to be violated are:

(a) Homogeneity of variance, which can be checked by a significance test such as Cochran's.
(b) Normal distribution of determination errors, which can be checked by graphical means.

9.7   Reproducibility

You will recall that in Chapter 3 we defined repeatability as 'The value below which the absolute difference between test results can be expected to lie with 95% confidence, when the results have been obtained under repeatability conditions'. By repeatability conditions we mean that the same operator, same apparatus, same laboratory have been used together with identical material. We calculated the repeatability (r) of a standard test method using:

$$r = 1.96\sqrt{(2)}\sigma$$

In Chapter 4, after discussing the t-test, we introduced the alternative formula:

$$r = t\sqrt{(2)}s$$

This second formula is more useful because it makes use of a sample standard deviation (s) whereas the first requires a population standard deviation (σ). Equivalent formulae for our present situation would be:

$$r = 1.96\sqrt{(2)}\sigma_w$$

and

$$r = t\sqrt{2}\sqrt{(\text{estimate of } \sigma_w^2)}$$

We have already obtained an estimate of the within-laboratory variance (σ_w^2) with 16 degrees of freedom. Substituting the estimate (0.021 25) and $t_{16} = 2.12$ we get:

$$r = 2.12\sqrt{(2)}\sqrt{(0.021\ 25)}$$
$$= 0.437$$

This result implies that, within any particular laboratory the absolute difference between any two determinations can be expected (with 95% confidence) to be less than 0.437. From the data in Table 9.2 we can calculate ten such differences for each laboratory and the largest of the forty differences is 0.4. It is comforting to observe that all the differences are less than the predicted upper limit and this inspires confidence that 95% of differences between future observations would be less than the repeatability (0.437).

Let us now turn our attention to the comparison of two determinations which wcre produced in *different* laboratories. For this we will need to use the concept of *reproducibility* which is defined in BS 5532: *Statistics – Vocabulary and Symbols* in two ways:

(1) *Qualitative definition.* The closeness of agreement between individual results obtained with the same method on identical test material but under different conditions (different operators, different apparatus, different laboratories and/or different times).

(2) *Quantitative definition.* The value below which the absolute difference between two single test results on identical material obtained by operators in different laboratories, using the standardized test method, may be expected to lie with 95% confidence.

To calculate a value for the reproducibility of a standard test we must take into account the *extra* variability that we expect to find in a set of determinations when they have been produced in two or more laboratories. The population variance for determinations made in a single laboratory is σ_w^2 but the population variance for results scattered around the thirty laboratories is $(\sigma_w^2 + \sigma_b^2)$. A simple modification to our repeatability formulae will give us a means of calculating the reproducibility (R).

$$R = 1.96\sqrt{(2)}\sqrt{(\sigma_w^2 + \sigma_b^2)}$$

and

$$R = t\sqrt{(2)}\sqrt{[(\text{estimate of } \sigma_w^2) + (\text{estimate of } \sigma_b^2)]}$$

We cannot use the first of the two formulae because we do not know the values of σ_w^2 and σ_b^2. We have an estimate of σ_w^2 with 16 degrees of freedom and we have an estimate of σ_b^2 with 3 degrees of freedom. Combining these degrees of freedom, as we did when calculating a confidence interval for thc true mean, gives approximately 7 degrees of freedom and a t value of 2.36. The reproducibility estimate is:

$$R = 2.36\sqrt{(2)}\sqrt{(0.021\,25 + 0.029)}$$
$$= 0.748$$

The value of the reproducibility (0.748) implies that there is a 95% chance that two determinations from different laboratories will have an absolute difference less than 0.748. From thc data in Table 9.2 we could calculate 150 such differences. The absolute values of these 150 differences are tabulated in the frequency distribution in Table 9.7.

We would expect to find 95% of these differences to be less than the value of the reproducibility estimate (0.748). In fact all 150 of the differences are less than this figure. It is not difficult to accept however that approximately 5% of the differences would have exceeded 0.748 if the determinations in Table 9.2 had been recorded with two decimal places rather than only one, and it is not unreasonable to suggest that we would get an absolute difference less than 0.748 if we obtained a new determination from each of two laboratories selected at random from the thirty laboratories within the company.

Table 9.7 Absolute differences between determinations from different laboratories

Absolute difference	0.0	0.1	0.2	0.3	0.4	0.5	0.6	Total
Number of pairs of determinations	13	29	31	31	21	13	12	150

9.8 Summary

The main concern of this chapter has been *reproducibility*. We have estimated the reproducibility of a particular test method for determining the concentration of ammonia in ammonia solution. This estimate is an upper limit for the difference we are likely to get between two determinations made in different laboratories.

When estimating reproducibility from the results of a precision experiment it is essential to take account of the additional errors that are introduced when determinations are made in different laboratories. We, therefore, made use of *analysis of variance* to separate the 'within-laboratory variation' from the 'between-laboratories variation'. This very powerful statistical technique will prove even more useful in later chapters when we consider more complex situations. Then we will require 'two-way analysis of variance' as distinct from the 'one-way analysis of variance' that we have used in this chapter.

Problems

(1) The data tabulated below has been extracted from Table 9.1. It contains the determinations made at level 4 (i.e. 25%) by the four laboratories.

Laboratory	A		B		C		D	
Determinations	24.8	25.1	24.7	25.1	25.6	25.3	25.4	25.0
	25.2	24.8	24.5	24.9	25.1	25.6	25.4	25.2
	25.1		24.8		25.4		25.0	

For ease of calculation 24.0 has been subtracted from the above results to give the table below:

Laboratory	A		B		C		D	
Determinations	0.8	1.1	0.7	1.1	1.6	1.3	1.4	1.0
	1.2	0.8	0.5	0.9	1.1	1.6	1.4	1.2
	1.1		0.8		1.4		1.0	
Mean	1.0		0.8		1.4		1.2	
SD	0.187		0.224		0.212		0.200	

(a) This part is aimed at the completion of the analysis of variance table given below:

Sources of variation	Sum of squares	Degrees of freedom	Mean square
Total			

   (i) Write the sources of variation into the table.
  (ii) Write in the degrees of freedom for the total.
 (iii) Write in the degrees of freedom for between-laboratories.
  (iv) By subtraction calculate the within-laboratories degrees of freedom.
   (v) Fill in the blanks in the table given below and then extract the sum of squares from this table and transfer to the analysis of variance table.
  (iv) Calculate the mean square for each row by dividing the sum of squares by the degrees of freedom and enter results into the analysis of variance table.

The analysis of variance table should now be complete.
The following questions all use results from the analysis of variance table.

(b) Calculate an estimate of within laboratories SD (σ_w) by taking the square root of the appropriate mean square.
(c) Calculate an estimate of between laboratories SD (σ_b).
(d) Calculate an estimate of repeatability using:

$$r = t\sqrt{(2)}\sqrt{(\text{estimate of } \sigma_w^2)}$$

(e) Calculate an estimate of reproducibility using:

$$R = t\sqrt{(2)}\sqrt{[(\text{estimate of } \sigma_w^2) + (\text{estimate of } \sigma_b^2)]}$$

(f) Before carrying out (b) and (d) we should carry out a test to determine that 'between-laboratories variance' is significant. Carry out this test as outlined in Chapter 9.
(g) The statistics calculated in (b) to (e) are only valid if the standard deviations within laboratories are not significantly different. Check whether this is true using Cochran's test.
(h) Carry out Dixon's test to check that the mean determination for each laboratory comes from the same normal distribution.
(i) Calculate a 95% confidence interval for the true mean determination and hence estimate the maximum bias for the method.
(j) Calculate a 95% confidence interval for Laboratory C and hence estimate its maximum bias.

1	2	3	4	5	6	7	8	9	10
Laboratory	Determination	Determination minus overall mean	Column 3 squared	Laboratory mean	Determination minus lab mean	Column 6 squared	Determination replaced by lab mean	Lab mean minus overall mean	Column 9 squared
A	0.8	-0.3	0.09		-0.2	0.04	1.0	-0.1	0.01
	1.1	0.0	0.00				1.0	-0.1	0.01
	1.2						1.0	-0.1	0.01
	0.8						1.0	-0.1	0.01
	1.1						1.0	-0.1	0.01
B	0.7	-0.4	0.16		-0.1	0.01		-0.3	0.09
	1.1	0.0	0.00	0.8	0.3	0.09			
	0.5				-0.3	0.09			
	0.9				0.1	0.01			
	0.8				0.0	0.00			
C	1.6	0.5	0.25						
	1.3	0.2	0.04						
	1.1								
	1.6								
	1.4								
D	1.4	0.1	0.01		0.2	0.04	1.2	0.1	0.01
	1.0	-0.1	0.01		-0.2	0.04	1.2	0.1	0.01
	1.4			1.2	0.2	0.04	1.2	0.1	0.01
	1.2				0.0	0.00	1.2	0.1	0.01
	1.0				-0.2	0.04	1.2	0.1	0.01
Total	22.0		Total SS			Within-lab SS			Between-labs SS
Mean		0.0							

(2) A number of criticisms can be made about the design of the experiment in Problem (1). Which of the following criticisms do you consider valid? Suggest designs which could overcome the valid criticisms.

(a) The determinations from Laboratory C should be ignored since four out of the six largest determinations came from this laboratory.

(b) The determinations from each laboratory should be made in more than one batch to give a true indication of the variability. In this experiment no indication was given to how many batches were used.

(c) Having five replicates is likely to result in the operator adjusting the determinations to reduce the variability.

(d) The standard deviations for each laboratory are different but only one estimate of σ_w is obtained.

─── 10 ───
Reproducibility (2)

10.1 Introduction

In Chapter 9 we calculated an estimate of the reproducibility of a particular test method. We made use of analysis of variance to subdivide the total variation in a set of determinations. This statistical technique enabled us to separate the within-laboratory variation from the between-laboratories variation as a prelude to estimating the between-laboratories variance (σ_b^2).

The technique we used in Chapter 9 would better be described as 'one-way analysis of variance' to distinguish it from the more powerful 'two-way analysis of variance' that we will use in this chapter. The extra power of the two-way technique is, unfortunately, only available to us if we are prepared to make extra assumptions. Fortunately these assumptions can be checked.

10.2 Analysing all four levels simultaneously

The data from the Chief Chemist's precision experiment was set out in Table 9.1 of the previous chapter. As we are now about to carry out further analysis of this data it is reproduced in Table 10.1.

The reader may recall that we used analysis of variance in the previous chapter to estimate the repeatability and reproducibility of the test method at the third level only. We could carry out a similar analysis to obtain precision estimates for the first, second and fourth levels. Indeed, we will do just that in a later chapter. First we will attempt to analyse the *whole* data set using a more sophisticated approach.

Table 10.1 contains four rows (one for each level) and four columns (one for each laboratory). Forgetting about levels and laboratories for the moment we can concentrate upon the sixteen groups or sixteen 'cells' as they are often called. Each cell contains several determinations and we can use analysis of variance, in exactly the same way that we did in the previous chapter, to break down the total variation into two components:

(a) Within cells.
(b) Between cells.

Table 10.1   Precision experiment with four levels

		Laboratory		
Level	A	B	C	D
1 (10%)	$d_1 = 10.0$ $d_2 = 10.0$ $d_3 = 10.1$ $d_4 = 9.9$ $d_5 = 10.0$	$d_6 = 9.9$ $d_7 = 10.0$ $d_8 = 10.0$ $d_9 = 10.1$ $d_{10} = 10.0$	$d_{11} = 10.1$ $d_{12} = 10.3$ $d_{13} = 10.0$ $d_{14} = 10.0$ $d_{15} = 10.1$	$d_{16} = 9.9$ $d_{17} = 10.0$ $d_{18} = 9.8$ $d_{19} = 9.9$ $d_{20} = 9.9$
2 (15%)	$d_{21} = 15.2$ $d_{22} = 14.9$ $d_{23} = 14.8$ $d_{24} = 15.1$ $d_{25} = 15.0$	$d_{26} = 15.0$ $d_{27} = 15.3$ $d_{28} = 15.3$ $d_{29} = 14.9$ $d_{30} = 15.0$	$d_{31} = 14.9$ $d_{32} = 15.1$ $d_{33} = 15.1$ $d_{34} = 15.2$ $d_{35} = 15.2$	$d_{36} = 14.8$ $d_{37} = 15.0$ $d_{38} = 14.7$ $d_{39} = 14.8$ $d_{40} = 14.7$
3 (20%)	$d_{41} = 19.7$ $d_{42} = 20.1$ $d_{43} = 19.7$ $d_{44} = 19.7$ $d_{45} = 19.8$	$d_{46} = 20.2$ $d_{47} = 20.3$ $d_{48} = 20.2$ $d_{49} = 20.0$ $d_{50} = 20.3$	$d_{51} = 20.3$ $d_{52} = 20.0$ $d_{53} = 19.9$ $d_{54} = 20.3$ $d_{55} = 20.0$	$d_{56} = 19.9$ $d_{57} = 19.8$ $d_{58} = 19.9$ $d_{59} = 19.9$ $d_{60} = 20.0$
4 (25%)	$d_{61} = 24.8$ $d_{62} = 25.1$ $d_{63} = 25.2$ $d_{64} = 24.8$ $d_{65} = 25.1$	$d_{66} = 24.7$ $d_{67} = 25.1$ $d_{68} = 24.5$ $d_{69} = 24.9$ $d_{70} = 24.8$	$d_{71} = 25.6$ $d_{72} = 25.3$ $d_{73} = 25.1$ $d_{74} = 25.6$ $d_{75} = 25.4$	$d_{76} = 25.4$ $d_{77} = 25.0$ $d_{78} = 25.4$ $d_{79} = 25.2$ $d_{80} = 25.0$

Rather than set out the calculations in tabular form (see Table 9.2) we will use the alternative formulae that are particularly suitable with a modern calculator. The means and standard deviations that we will need are set out in Table 10.2.

Table 10.2   Cell means and standard deviations

		Laboratory			
Level		A	B	C	D
1	Mean SD	10.0 0.071	10.0 0.071	10.1 0.122	9.9 0.071
2	Mean SD	15.0 0.158	15.1 0.187	15.1 0.122	14.8 0.122
3	Mean SD	19.8 0.173	20.2 0.122	20.1 0.187	19.9 0.071
4	Mean SD	25.0 0.187	24.8 0.224	25.4 0.212	25.2 0.200

total sum of squares = (degrees of freedom) (SD of all 80
$$\text{determinations})^2$$
$$= 79(5.663\,060\,406)^2$$
$$= 2533.55$$

within-cells sum of squares = $\sum$[(degrees of freedom) (SD of 5 determinations
in cell)^2]
$$= [4(0.071)^2 + 4(0.071)^2 + 4(0.122)^2 + 4(0.071)^2$$
$$+ 4(0.158)^2 + 4(0.187)^2 + 4(0.122)^2 + 4(0.122)^2$$
$$+ 4(0.173)^2 + 4(0.122)^2 + 4(0.187)^2 + 4(0.071)^2$$
$$+ 4(0.187)^2 + 4(0.224)^2 + 4(0.212)^2 + 4(0.200)^2]$$
$$= 1.500$$

between-cells sum of squares = (degrees of freedom) (SD of the 16 cell means)^2
$$\times \text{ (no. of observations per cell)}$$
$$= (15)(5.810\,393)^2(5)$$
$$= 2532.050$$

Premature rounding of numbers can have a disastrous effect on the above calculations. You are, therefore, advised to carry all the decimal places that the pocket calculator gives you and only round off the final results. These are summarized in the analysis-of-variance table (10.3).

Table 10.3   One-way analysis of variance on the whole of Table *10.1*

Source of variation	Sum of squares	Degrees of freedom	Mean square
Between cells	2532.050	15	168.8033
Within cells	1.500	64	0.023 438
Total	2533.550	79	

We could carry out an F-test on the mean squares in the above table. If we did so we would find that the between-cells mean square was very, very significantly greater than the within cell mean square. What practical conclusions could we draw from this finding? Unfortunately, the F-test would only confirm what we must already know. Commonsense could have told us that some of the variability in Table 10.1 results from using four laboratories in the experiment, but a great deal more is due to using four different levels of concentration. This is confirmed by the table of means (Table 10.4).

A very high percentage of the variation in Table 10.1 results from the use of four levels and the presence of this 'between levels' variation tends to obscure the much smaller variation between laboratories. Fortunately, we can eliminate the between-levels variation in Table 10.2 by subtracting each 'level mean' from the four cell means at that level. This has been done in Table 10.5.

Table 10.4   Level means

Level	1	2	3	4	Mean of level means	SD of level means
Mean	10.00	15.00	20.00	25.10	17.525	6.493 779

Table 10.5   Cell mean minus level mean – revealing bias

| | Laboratory | | | | | |
Level	A	B	C	D	Mean	SD
1	0.00	0.00	0.10	−0.10	0.000	0.0816
2	0.00	0.10	0.10	−0.20	0.000	0.1414
3	−0.20	0.20	0.10	−0.10	0.000	0.1826
4	−0.10	−0.30	0.30	0.10	0.000	0.2582
Mean	−0.075	0.000	0.150	−0.075	0.000	

Having eliminated the variation between levels in Table 10.5 we can now see evidence of bias amongst three of the four laboratories. Laboratory C appears to have a positive bias of 0.150, whilst laboratories A and D appear to have a negative bias of 0.075. We will have more to say about this bias later. Let us now return to our main theme of quantifying variability in order to estimate the precision of the test method. To isolate the variation between laboratories we need the laboratory means which are set out in Table 10.6.

Table 10.6   Laboratory means

| | Laboratory | | | | Mean of lab means | SD of lab means |
	A	B	C	D		
Mean	17.45	17.525	17.675	17.45	17.525	0.106066

Using the laboratory means in Table 10.6 and the level means in Table 10.4 we can quantify the variation between laboratories and the variation between levels. In doing so we will be using *two-way* analysis of variance, but this is only a simple extension of what we have done already.

10.3 Two-way analysis of variance

When using two-way analysis of variance we classify, or categorize, the data in *two* ways. With the data in Table 10.1 the two categories are 'levels' and 'laboratories'. Using the laboratory means in Table 10.6 we can calculate the between-laboratories sum of squares:

between-laboratories sum of squares = (degrees of freedom) (SD of the 4 laboratory mean)^2 × (no. of determinations
per lab)

$$= (3) (0.106\,066)^2 (20)$$
$$= 0.675$$

Using the level means in a similar way we can calculate the between-levels sum of squares:

between-levels sum of squares = (degrees of freedom) (SD of the 4 level
means)^2 × (no. of determinations per level)

$$= (3) (6.493\,779)^2 (20)$$
$$= 2530.150$$

We can see from the sums of squares just calculated that the variation between laboratories is very much less than the variation between levels; indeed, the latter accounts for a considerable proportion of the total variation. Adding together the between-laboratories sum of squares (0.675) and the between-levels sum of squares (2530.150) gives 2530.825. We see that this total is less than the between-cells sum of squares in Table 10.3. Obviously the variability between laboratories plus the variability between levels does *not* account for *all* of the variability between cells. A third component is necessary and this is referred to as 'interaction' in Table 10.7.

Table 10.7 Breaking down the between-cells variation

Source of variation	Sum of squares	Degrees of freedom	Mean square
Between levels	2530.150	3	843.383
Between labs	0.675	3	0.225
Interaction	1.225	9	0.136
Between cells	2532.050	15	
Within cells	1.500	64	0.023 437
Total	2533.550	79	

Before we explain what is meant by the very important word 'interaction', several comments need to be made concerning the above table:

(a) Between-levels has three degrees of freedom because we have used four levels in the experiment.
(b) Between-laboratories has three degrees of freedom because we have used four laboratories in the experiment.
(c) Interaction degrees of freedom is equal to between-levels degrees of freedom multiplied by between-laboratories degrees of freedom.
(d) The dotted line across the table implies that the sum of squares immediately below the line is equal to the sum of the entries above the line, i.e. $2532.050 = 2530.150 + 0.675 + 1.225$.
(e) The dotted line across the table implies that the degrees of freedom immediately below the line is equal to the sum of the entries above the line, i.e. $15 = 3 + 3 + 9$.
(f) There is some duplication in Table 10.7. We could remove the 'between cells' row from the table to obtain the conventional two-way analysis-of-variance table (Table 10.8). Note that the within-cells entry has been renamed 'residual'.

Table 10.8   Two-way analysis of variance table

Source of variation	Sum of squares	Degrees of freedom	Mean square
Between levels	2530.150	3	843.3833
Between laboratories	0.675	3	0.2250
Interaction	1.225	9	0.1361
Residual	1.500	64	0.023437
Total	2533.550	79	

Several *F*-tests could be carried out using the mean squares in Table 10.8. It is important to start at the bottom of the table and work our way upwards. *First* we compare the interaction mean square with the residual mean square and find that the former is significantly greater than the latter. This tells us that there is variation between the sixteen cells of Table 10.1 (over and above the lab-to-lab variation and the level-to-level variation) that is greater than we would expect in the light of the within-cells variation. The full implication of this interaction will be discussed when we have completed the *F*-tests.

Next we compare the between-laboratories mean square with the interaction mean square. We find that the former is not significantly greater than the latter. (Note that it is not valid to compare the between-laboratories mean square with the residual mean square now that we have shown the interaction to be significant.) The lack of significance in this second comparison suggests that there is no consistent lab-to-lab variation operating equally at the four levels. The significance of the interaction, on the other hand, implies that there is a lab-to-lab variation but it is not the same at each level. Perhaps this point will become clear if we compare our data with a hypothetical set of data that might have arisen if we had used four other laboratories. The two sets of data are illustrated in Fig. 10.1. The actual data from our experiment is shown in Fig. 10.1(a) alongside some hypothetical data which contains no evidence of interaction. For this hypothetical data the interaction sum of squares would be zero.

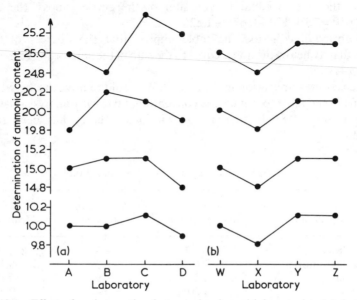

Figure 10.1   Effect of an interaction between levels and laboratories. (a) Actual data, (b) Hypothetical data

Turning first to the hypothetical data we see that laboratory X has a negative bias and that this bias has the same magnitude at every level. Laboratories Y and Z, on the other hand, have a positive bias; but again the bias is constant and does not change from level to level. Turning now to the data from our experiment we see that Laboratory C has a positive bias, but the magnitude of this bias is *not* constant. Similar statements could be made about laboratories A, B and D. The fundamental difference between the hypothetical data and the actual data is further emphasized if we compare Table 10.9 with Table 10.5 considered earlier.

We can see that the consistency which is to be found in the columns of Table 10.9 is not apparent in Table 10.5 reflecting what we have already observed in Fig. 10.1. Perhaps Fig. 10.1(a) can best be described by saying that the four laboratories exhibit a *variable bias*, in contrast to the *fixed bias* shown by the

Table 10.9 Cell mean minus level mean (hypothetical data from Figure 10.1(b)).

| | Laboratory | | | | |
Level	W	X	Y	Z	Mean
1	0.000	−0.200	0.100	0.100	0.000
2	0.000	−0.200	0.100	0.100	0.000
3	0.000	−0.200	0.100	0.100	0.000
4	0.000	−0.200	0.100	0.100	0.000
Mean	0.000	−0.200	0.100	0.100	0.000

hypothetical laboratories in Fig. 10.1(b). Furthermore, the statistical significance of the interaction in Table 10.8 suggests that the variability in bias is not simply a 'chance effect', but would be very likely to arise again if the experiment were repeated.

The variation in bias in Fig. 10.1(a) appears to be random. This impression is confirmed by an inspection of the 'deviations' in Table 10.10, which also appear to be random. These deviations were calculated from the formula:

$$\text{deviation} = \text{cell mean} - \text{lab mean} - \text{level mean} + \text{overall mean}$$

and therefore contain *no* variability between levels and *no* variability between laboratories.

Table 10.10 Deviations (Actual data)

| | Laboratory | | | | |
Level	A	B	C	D	Mean
1	0.075	0.000	−0.050	−0.025	0.000
2	0.075	0.100	−0.050	−0.125	0.000
3	−0.125	0.200	−0.050	−0.025	0.000
4	−0.025	−0.300	0.150	0.175	0.000
Mean	0.000	0.000	0.000	0.000	0.000

Because we have eliminated the variation between levels and the variation between laboratories, the means in Table 10.10 are all equal to zero. Because there is a significant interaction between levels and laboratories, however, the deviations in Table 10.10 cannot be dismissed as the result of within-laboratory error. If a similar table were drawn up for the hypothetical data we would find that all the deviations were equal to zero.

We have expressed the opinion that the deviations in Table 10.10 are random in nature. With only four levels and only four laboratories it is difficult to either refute or support such an assertion, of course. If, however, a clear pattern were discernible in the deviations this would require further investigation and might help us to decide *why* the interaction existed. Suppose, for example, we found larger deviations at the higher levels; this would indicate that the laboratory biases were proportional rather than fixed. Ideally such a discovery would not occur during an inter-laboratory trial, because the error structure should have been investigated previously. Furthermore, the discovery of proportional bias would indicate that an assumption was being violated. We will now turn our attention to the many assumptions underlying two-way analysis of variance.

10.4   Assumptions underlying two-way analysis

In the previous chapter we introduced a model:

$$y = \mu + \beta + \delta$$

which facilitated the discussion of the assumptions underlying one-way analysis of variance. A suitable model for two-way analysis of variance would be:

$$y = \mu + \beta + \alpha + \delta$$

where μ = the true mean determination for a particular level of concentration;
β = the true mean bias of a particular laboratory;
α = the deviation of a particular laboratory from its true mean bias, at a particular level;
δ = random testing error.

If we carry out a precision experiment using four levels then μ in the model will have four values and these can be estimated by the level means in Table 10.4. Thus μ_1 can be estimated by m_1 which is equal to 10.0, μ_2 can be estimated by m_2 which is equal to 15.0, etc. If we involve four laboratories in the experiment then β will have four values which can be estimated by the laboratory means in the bottom row of Table 10.5. Thus β_A can be estimated by b_A, β_B by b_B, etc. The sixteen laboratory-level combinations give rise to sixteen values of α in the model and these in turn can be estimated by the sixteen deviations in Table 10.10. Thus α_1 can be estimated by a_1 which is equal to 0.075, α_2 can be estimated by a_2 which is equal to 0.000, etc. If we obtain five determinations at each laboratory-level combination then we have eighty determinations in total, as in Fig. 10.1. Each determination contains a random testing error and we have, therefore, eighty values of δ in the model. These can be estimated by the eighty residuals which are calculated from:

$$\text{Residual} = \text{determination} - \text{cell mean}$$

The estimates in Table 10.11 show us how each of the eighty determinations can

Table 10.11   Estimates of the terms in the model

		Laboratory			
Level		A $b_A = -0.075$	B $b_B = 0.000$	C $b_C = 0.150$	D $b_D = -0.075$
1	$m_1 = 10.00$	$a_1 = 0.075$ $e_1 = 0.00$ $e_2 = 0.00$ $e_3 = 0.10$ $e_4 = -0.10$ $e_5 = 0.00$	$a_2 = 0.000$ $e_6 = -0.10$ $e_7 = 0.00$ $e_8 = 0.00$ $e_9 = 0.10$ $e_{10} = 0.00$	$a_3 = -0.050$ $e_{11} = 0.00$ $e_{12} = 0.20$ $e_{13} = -0.10$ $e_{14} = -0.10$ $e_{15} = 0.00$	$a_4 = 0.025$ $e_{16} = 0.00$ $e_{17} = 0.10$ $e_{18} = -0.10$ $e_{19} = 0.00$ $e_{20} = 0.00$
2	$m_2 = 15.00$	$a_5 = 0.075$ $e_{21} = 0.20$ $e_{22} = -0.10$ $e_{23} = -0.20$ $e_{24} = 0.10$ $e_{25} = 0.00$	$a_6 = 0.100$ $e_{26} = -0.10$ $e_{27} = 0.20$ $e_{28} = 0.20$ $e_{29} = -0.20$ $e_{30} = -0.10$	$a_7 = -0.050$ $e_{31} = -0.20$ $e_{32} = 0.00$ $e_{33} = 0.00$ $e_{34} = 0.10$ $e_{35} = 0.10$	$a_8 = -0.125$ $e_{36} = 0.00$ $e_{37} = 0.20$ $e_{38} = -0.10$ $e_{39} = 0.00$ $e_{40} = -0.10$
3	$m_3 = 20.00$	$a_9 = -0.125$ $e_{41} = -0.10$ $e_{42} = 0.30$ $e_{43} = -0.10$ $e_{44} = -0.10$ $e_{45} = 0.00$	$a_{10} = 0.200$ $e_{46} = 0.00$ $e_{47} = 0.10$ $e_{48} = 0.00$ $e_{49} = -0.20$ $e_{50} = 0.10$	$a_{11} = 0.050$ $e_{51} = 0.20$ $e_{52} = -0.10$ $e_{53} = -0.20$ $e_{54} = 0.20$ $e_{55} = -0.10$	$a_{12} = -0.025$ $e_{56} = 0.00$ $e_{57} = -0.10$ $e_{58} = 0.00$ $e_{59} = 0.00$ $e_{60} = 0.10$
4	$m_4 = 25.10$	$a_{13} = -0.025$ $e_{61} = -0.20$ $e_{62} = 0.10$ $e_{63} = 0.20$ $e_{64} = -0.20$ $e_{65} = 0.10$	$a_{14} = -0.300$ $e_{66} = -0.10$ $e_{67} = 0.30$ $e_{68} = -0.30$ $e_{69} = 0.10$ $e_{70} = 0.00$	$a_{15} = 0.150$ $e_{71} = 0.20$ $e_{72} = -0.10$ $e_{73} = -0.30$ $e_{74} = 0.20$ $e_{75} = 0.00$	$a_{16} = 0.175$ $e_{76} = 0.20$ $e_{77} = -0.20$ $e_{78} = 0.20$ $e_{79} = 0.00$ $e_{80} = -0.20$

be broken down into its constituent parts. Perhaps this will be clarified if we 'reconstruct' the first and the last determinations.

The model tells us that:

$$d_1 = \mu_1 + \beta_A + \alpha_1 + \delta_1$$

Replacing the population unknowns by their estimates gives us:

$$d_1 = m_1 + b_A + a_1 + e_1$$
$$= 10.00 + (-0.075) + 0.075 + 0.000$$
$$= 10.000$$

Similarly

$$d_{80} = m_4 + b_D + a_{16} + e_{80}$$
$$= 25.10 + (-0.075) + 0.175 + (-0.200)$$
$$= 25.000$$

The 'reconstructed' determinations are in perfect agreement with the actual determinations in Table 10.1. 'So 25.0 = 25.0' the reader might say, 'What is the point of this arithmetic?' The purpose of producing Table 10.11 is to help the reader to appreciate the implications of the assumptions underlying two-way analysis of variance. These assumptions are:

(a) The within-laboratory variance (σ_w^2) is constant, i.e. all laboratories are equally variable at every level. In other words all sixteen cells in Table 10.1 would have the same standard deviation if we had continued testing indefinitely rather than stopping after five determinations. If the precision of the test method were dependent on the level of concentration this assumption would not be satisfied, but this point could be checked by carrying out a Cochran's test on the sixteen cell standard deviations in Table 10.2.

(b) The between-laboratories variance (σ_b^2) is constant, i.e. laboratories are not more variable at one level than at another level. If laboratories exhibited proportional bias it is unlikely that this condition would be satisfied. A Cochran's test could be carried out, using the four row standard deviations in Table 10.5, in order to check this assumption.

(c) For the F-tests to be valid on the mean squares in Table 10.8 the δ values, the β values and the α values in the model must have normal distributions. This assumption could be checked by carrying out Dixon's tests on:

  (i) the largest and smallest determinations in each of the sixteen cells of Table 10.1;

  (ii) the largest and smallest laboratory means in Table 10.6;

  (iii) the largest and smallest deviations in Table 10.10.

Obviously there is a lot of work involved in carrying out the outlier tests to check the assumptions underlying the two-way analysis of variance. Failure to carry out the tests, however, may result in the application of analysis of variance to data which is quite unsuitable and hence lead to the drawing of invalid conclusions or the calculation of biased estimates.

Application of Cochran's test to the sixteen cell standard deviations of Table 10.2 fails to demonstrate that the largest variance $[(0.224)^2]$ is significantly greater than the other variances. (Test statistic $= 0.134$, critical value $= 0.230$ for 5% significance.) Thus assumption (a) appears to be satisfied.

Application of Cochran's test to the four row standard deviations of Table 10.5 fails to demonstrate that the largest variance $[(0.2582)^2]$ is significantly greater than the other variances. (Test statistic $= 0.526$, critical value $= 0.684$ for 5% significance.)

Application of Dixon's test to each of the sixteen cells of Table 10.1 gives a significant result in only one cell. Amongst the five determinations made by laboratory A on the 20% solution the 20.1 can be regarded as an outlier. (Test statistic $= 0.75$, critical values are 0.710 at 5% and 0.821 at 1% significance.) Perhaps it is unwise, however, to use a 5% significance level when carrying out a succession of sixteen tests.

Application of Dixon's test to the laboratory means in Table 10.6 fails to detect an outlier. (Test statistic $=0.667$, critical value $=0.829$ for 5% significance.) Application of Dixon's test to the sixteen deviations in Table 10.10 is equally unrewarding. (Test statistic $=0.389$, critical value $=0.546$ for 5% significance.)

The end-result of this outlier testing must be the conclusion that Dixon's and Cochran's tests have failed to reveal any gross violations of the assumptions underlying two-way analysis of variance. The perceptive reader may, however, have already drawn the *opposite* conclusion. There is a strong indication within the standard deviations of Table 10.2 that the precision of the test method is not constant. The standard deviations appear to increase as the concentration increases, but Cochran's test did not respond to this trend. In fairness to Cochran and to Dixon it should be pointed out that their tests were never intended to detect gradual change, but were designed to detect gross deviations from a distinct clustering. Whilst the outlier tests will help us to identify certain violations of assumptions, they are not all-powerful.

In a later chapter we will examine the recommendations of BS 5497 and IS 4259 which both offer guidance on the conduct of precision experiments. We will pay particular attention to the checks of assumptions that are prescribed in these two documents and we will apply these checks to the data in Table 10.1. For the moment let us forget about assumptions and see how the results of a two-way analysis of variance can be used to estimate the reproducibility of a test method.

10.5   Estimating reproducibility

When F-tests were carried out on the mean squares in Table 10.8 it was revealed that the interaction mean square was significantly greater than the residual mean square. The significance of the interaction has implications when we come to estimate the reproducibility of the test method. To obtain precision estimates from the two-way analysis of variance we will use:

> Repeatability estimate $= t\sqrt{(2)}s_w$
>
> Reproducibility estimate $= t\sqrt{[2(s_w^2 + s_I^2 + s_b^2)]}$
>
> where s_w^2 is an estimate of the within-laboratory variance (σ_w^2)
>
> s_I^2 is an estimate of the interaction variance (σ_I^2)
>
> s_b^2 is an estimate of the between-laboratories variance (σ_b^2)

The three variances, σ_w^2, σ_I^2 and σ_b^2 are not known, of course, but can be estimated from Table 10.8 as follows:

$s_w^2 = $ residual mean square

$\quad = 0.023437$

$s_I^2 =$ (interaction mean square – residual mean square)/(no. of determinations in each cell)

$$= (0.1361 - 0.023\,437)/5$$

$$= 0.022\,533$$

$s_b^2 =$ (between-laboratories mean square – interaction mean square)/(no. of determinations in each laboratory)

$$= (0.2250 - 0.1361)/20$$

$$= 0.004\,445$$

Using the variance estimates we can now calculate estimates of repeatability and reproducibility for the test method. To obtain the t values we must use an appropriate number of degrees of freedom. As the residual mean square has 64 degrees of freedom we will use $t_{64} = 2.00$ when calculating the repeatability estimate:

$$\text{Repeatability estimate} = t_{64}\sqrt{(2)}s_w$$

$$= 2.00\sqrt{(2)}\sqrt{(0.023\,437)}$$

$$= 0.433$$

The interaction mean square has nine degrees of freedom and the between-laboratories mean square has three degrees of freedom. How many degrees of freedom should we attribute to the combination of s_w^2, s_I^2 and s_b^2 that is used in the calculation of a reproducibility estimate? For this combination of variances the degrees of freedom can be calculated using a rather complex formula given in IS 4259. The result of the calculation is approximately thirty degrees of freedom and $t_{30} = 2.04$.

$$\text{Reproducibility estimate} = 2.04\sqrt{(2)}\sqrt{(0.023\,437 + 0.022\,533 + 0.004\,445)}$$

$$= 0.647$$

If all the assumptions underlying two-way analysis of variance were satisfied by the data in Table 10.1 then the precision estimates we have just calculated would be applicable to a range of concentrations from 10% to 25%. The repeatability estimate implies that two determinations made in the *same laboratory* on the same sample are unlikely to differ by more than 0.433. The reproducibility estimate suggests that the difference could be larger, up to 0.647, if the two determinations were made on identical samples but in *different* laboratories.

As we earlier expressed reservations concerning the assumption of constant precision we must also reserve judgement on the repeatability and reproducibility estimates. In a later chapter we will show that the precision is related to concentration, then we will obtain different precision estimates for different levels. The medium-range (i.e. $17\frac{1}{2}\%$) estimates will be very similar to those we have just calculated.

10.6 Summary

In this chapter we have concentrated on the very powerful statistical technique *two-way analysis of variance*. We used it to break down the total variation in a set of determinations into four components:

(a) Between laboratories
(b) Between levels
(c) Interaction between laboratories and levels
(d) Residual.

Each of these components tells us something about the bias and/or precision of the test method and we used the mean squares from the analysis of variance to calculate estimates of repeatability and reproducibility.

We also discussed the assumptions underlying analysis of variance and we used Dixon's test and Cochran's test to check these assumptions. This checking was not entirely satisfactory in that it failed to reveal a rather obvious violation of the assumption that 'the precision of the test method is constant'. We will have more to say about this assumption in Chapter 12.

Problems

(1) A laboratory uses HPLC to assay the concentration of dioxamine in a liquid. The laboratory manager has for some time been worried about the precision of the determinations and has focused his attention on two likely causes of the variability – due to operators and due to a day-to-day variation. He accordingly designs a precision experiment in which the same sample is replicated thrice by three operators on four different days. The results of the precision experiment are given below together with summary statistics.

| | Days | | | | | | | |
Operators	A		B		C		D	
W	28.1		26.9		26.9		26.9	
	27.7	28.8	27.0	26.8	27.9	27.4	26.9	27.5
X	26.6		26.1		25.9		26.3	
	27.2	26.9	25.8	26.7	25.9	26.5	26.5	26.4
Z	28.1		27.0		27.2		27.5	
	28.4	28.4	27.5	27.4	26.6	26.9	27.2	27.8

Operators		Days				Mean
		A	B	C	D	
W	Mean	28.2	26.9	27.4	27.1	27.4
	SD	0.556	0.100	0.500	0.346	
X	Mean	26.9	26.2	26.1	26.4	26.4
	SD	0.300	0.458	0.346	0.100	
Z	Mean	28.3	27.3	26.9	27.5	27.5
	SD	0.173	0.265	0.300	0.300	
Mean		27.8	26.8	26.8	27.0	27.1

$$\text{SD of all 36 determinations} = 0.734\,458\,012$$
$$\text{Total sum of squares} = 35(0.734\,458\,012)^2 = 18.88$$

(a) Calculate the standard deviation of the cell means and hence complete the following table:

Source of variation	Sum of squares	Degrees of freedom	Mean square
Between-cells			
Within-cell			
Total	18.88		

(b) Calculate the standard deviation of the day means and of the operator means and hence complete an extended analysis-of-variance table.

(c) Determine using significance tests whether there is a fixed or variable operator bias and whether there is a day-to-day variation in bias.

(d) Assuming that the operators and days are representative samples of the populations, calculate the standard deviation which is applicable to a single determination of similar magnitude to those in the precision experiment.

(e) If the laboratory decide to carry out duplicate readings, how should these be scheduled to give the best estimate of true mean level of dioxamine and by how much will the readings reduce the standard deviation of the mean of the two determinations?

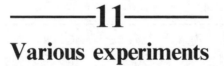

Various experiments

11.1 Introduction

In the previous chapter we made use of two-way analysis of variance to analyse the results of a precision experiment. This powerful technique enabled us to separate the variation between levels from the variation between laboratories, but it also thrust upon us the concept of an interaction between laboratories and levels.

Like all other powerful statistical techniques, analysis of variance has underlying assumptions which must be satisfied if we are to have confidence in the conclusions or estimates. We discussed these assumptions, making reference to a model, and we used outlier tests to check the assumptions.

In this chapter we will make use of analysis of variance in a variety of situations. We will examine four experiments, all of which differ in some respects from the precision experiment in Chapter 10. By comparing the subtle differences between the experiments we will, hopefully, reveal the power and versatility of this statistical technique. To reduce the tedium of what could be a very long chapter we will not get involved in data analysis, but will concentrate on setting up the analysis of variance table and inserting the degrees of freedom.

11.2 Experiment One

We wish to select a test method to determine the concentration of a particular impurity in a final product. *Four* test methods are available and it is considered desirable to evaluate each method with each of the **three** physical forms of the product – paste, powder and gel. Eight samples of each form are prepared; all 24 samples containing 2% of the impurity. Two samples of each form are allocated to each of the four methods but, due to a misunderstanding, only one sample is analysed by method B and three samples are analysed by method D. The results of the experiment are shown in Table 11.1.

In Table 11.1 we have twelve cells with either one, two or three determinations in each. In some respects Table 11.1 is similar to Table 10.1 which had sixteen cells with five determinations in each. The rows in the new table represent

Table 11.1   Results of Experiment One

Physical form	Method			
	A	*B*	*C*	*D*
Paste	d_1 d_2	d_3	d_4 d_5	d_6 d_7 d_8
Powder	d_9 d_{10}	d_{11}	d_{12} d_{13}	d_{14} d_{15} d_{16}
Gel	d_{17} d_{18}	d_{19}	d_{20} d_{21}	d_{22} d_{23} d_{24}

different physical forms, in contrast to the different levels of concentration considered earlier. In the previous chapter we had one column for each laboratory, whereas we now have one column for each method. Despite the similarity of the two tables there is one very important difference between the two experiments. This difference is centred upon the way in which the two experiments were designed:

(a) Whereas the four laboratories in Chapter 10 were a random sample from a population of thirty laboratories, the four methods here are the *only* methods available. Each method is important in its own right, rather than as a representative of a wider group.

(b) Whereas the four levels in Chapter 10 were representative of a whole range of concentration (though certainly not a random sample), the three physical forms here are the *only* forms in which the product is manufactured.

Despite this difference we can apply two-way analysis of variance just as well to this new situation as to the one in Chapter 10. We cannot calculate the sums of squares, of course, because we do not have the actual determinations, but we can complete two columns of the 'analysis-of-variance table (Table 11.2).

'Between Methods' has three degrees of freedom because we have used four methods. 'Between forms' has two degrees of freedom because we have used three physical forms. To obtain the 'interaction' degrees of freedom we multiply these two numbers. $3 \times 2 = 6$. The number of degrees of freedom *within* any cell of Table 11.1 will be one less than the number of determinations in the cell. The

Table 11.2   Analysis of variance for Experiment One

Source of variation	Degrees of freedom
Between methods	3
Between forms	2
Interaction	6
Residual	12
Total	23

three cells containing only one determination will each have zero degrees of freedom, whilst the six cells containing two determinations will each have one degree of freedom, etc. Adding the degrees of freedom within the twelve cells gives us twelve degrees of freedom for the 'residual'.

To calculate the sums of squares for Table 11.2 we could use exactly the same formulae that we used in the previous chapter. The mean squares could then be calculated in the usual way. The whole analysis-of-variance table is therefore in keeping with the two-way analysis of variance carried out in Chapter 10. When we come to the next step in the analysis, however, we find differences.

The between-methods mean square and the between-forms mean square must both be tested against the residual mean square and *not* against the interaction, as we did in the previous chapter. The reason for this change lies in the difference referred to earlier, namely the four methods do not constitute a sample from a larger group of methods, about which we wish to draw conclusions. Though we can use the same model for each of the two experiments it must be interpreted rather differently in each case. In Chapter 10 we spoke of the 'between-laboratories variance' and the 'interaction variance'; but in the present context it would be meaningless to speak of 'between methods variance' or 'interaction variance' or 'between-forms variance'. Many statistics texts would explain this difference by saying that the present situation calls for a 'parametric model' whilst the situation in Table 10 required a 'random model'. We are not happy with these descriptions, largely because the word 'parametric' has many different meanings in different branches of science and technology. We prefer to use *complete model* for experiments in which all possibilities have been included, such as the three physical forms and the four test methods in Experiment One. For experiments in which the explored possibilities have been selected at random from a large population, we prefer the name *sampling model*.

Consider the model we used in the previous chapter:

$$d = \mu + \beta + \alpha + \delta$$

By interpreting this model in two rather different ways we can make it fit both experiments. The two interpretations are set alongside each other in Table 11.3.

The model we have used is rather abstract and the reader may have found the comparison of the two interpretations rather tedious. Let us close this discussion, therefore, by emphasizing the practical implications of the similarities and the differences of the two experiments (i.e. the precision experiment of Chapter 10 and Experiment One of this chapter):

(a) To calculate the sums of squares, degrees of freedom and mean squares we follow exactly the same rules for both experiments.
(b) When using the *complete model* in Experiment One we test each of the mean squares against the residual mean square.
(c) When using the *sampling model* in Chapter 10 we first test the interaction

Table 11.3 Interpretation of the model $d = \mu + \beta + \alpha + \delta$

Term in model	A sampling model for the precision experiment in Chapter 10			A complete model for Experiment One of Chapter 11		
	No. of values within the experiment	No. of values within the population	Interpretation	No. of values within the experiment	No. of values within the population	Interpretation
μ	4	∞	The true mean determination at a particular level of concentration.	3	3	The true mean determination for a particular physical form.
β	4	30 but assumed to be ∞	Deviation from μ due to consistent bias of a particular laboratory. The variance of the β values for the whole population of laboratories is known as the 'between laboratories variance' (σ_b^2).	4	4	Deviation from μ due to consistent bias of a particular method.
α	16	∞	Deviation from $(\mu+\beta)$ due to fluctuation in bias at a particular laboratory-level combination. The variance of the α values for the whole population of laboratory-level combinations is known as the 'interaction variance' (σ_i^2).	12	12	Deviation from $(\mu+\beta)$ due to fluctuation in bias at a particular method-form combination.
δ	80	∞	Deviation from $(\mu+\beta+\alpha)$ of a particular determination. The variance of the δ values for all possible determinations that could have been made at a particular laboratory-level combination is known as the 'testing variance' (σ^2).	24	∞	Deviation from $(\mu+\beta+\alpha)$ of a particular determination. The variance of the δ values for all possible determinations that could have been made at a particular method-form combination is known as the 'testing variance' (σ^2).

mean square against the residual mean square and if this is significant we then test the other mean squares against the interaction mean square.

(d) When using the *sampling model* we draw conclusions about the whole population though we have only involved a sample in the experiment (e.g. we calculate a reproducibility estimate that is applicable to *any* pair of laboratories not just the ones in the experiment).

(e) When using the *complete model* we confine our conclusions to the items (e.g. methods, physical forms, etc.) included in the experiment. For example, a significant interaction might indicate that a specific test method was biased only when testing one particular physical form. Furthermore, we would almost certainly wish to follow the analysis of variance with several more precise comparisons. We might for example, compare the 'best' test method with the mean of the others or we might single out one physical form for special consideration.

11.3 Experiment Two

As in Experiment One we wish to compare four test methods. We are not however concerned with three physical forms of the product, but with the performance of the many operators who could be called upon to carry out the test. From the unlimited number of operators who *might* be involved we select a random sample of five and ask each operator to make two determinations on a standard solution using each of the four methods. The results of the experiment are given in Table 11.4.

As with Experiment One we can apply two-way analysis of variance to the data in Table 11.4. Sums of squares could be calculated for 'between-methods', 'between-operators', 'interaction' and, of course, 'residual'. The appropriate degrees of freedom are given in Table 11.5.

Again we can use the model equation that has served us well on two occasions already. We must, of course, match the interpretation of the model to the design of the experiment and the reader may already have sensed that this experiment appears to be somewhere between the two experiments that we have just

Table 11.4 Results of Experiment Two

Operator	Method							
	A		B		C		D	
1	d_1	d_2	d_3	d_4	d_5	d_6	d_7	d_8
2	d_9	d_{10}	d_{11}	d_{12}	d_{13}	d_{14}	d_{15}	d_{16}
3	d_{17}	d_{18}	d_{19}	d_{20}	d_{21}	d_{22}	d_{23}	d_{24}
4	d_{25}	d_{26}	d_{27}	d_{28}	d_{29}	d_{30}	d_{31}	d_{32}
5	d_{33}	d_{34}	d_{35}	d_{36}	d_{37}	d_{38}	d_{39}	d_{40}

Table 11.5   Analysis of variance for Experiment Two

Source of variation	Degrees of freedom
Between-methods	3
Between-operators	4
Interaction	12
Residual	20
Total	39

discussed. 'Methods' calls for a *complete* model, whereas 'operators' calls for a *sampling* model. In fact this is a situation which demands a *mixed model* in which the values of β (representing methods) are complete whilst the values of μ (representing operators) are sampled from a population.

The practical implication of using the mixed model is that the interaction mean square and the between-methods mean square must be tested against the residual mean square. If the interaction is found to be significant, however, we must test the between-operators mean square against the interaction mean square.

11.4   Experiment Three

An analyst wishes to check the bias and precision of a new test method. This automatic method may be used in as many as sixty laboratories throughout the company to analyse samples in large batches. The analyst selects three laboratories from those already using the method and sends six identical samples to each, with instructions that two samples are to be inserted at random in each of three separate batches.

With three laboratories each making two determinations in each of three batches the analyst will obtain ($3 \times 3 \times 2 = 18$) determinations in total. How are these results to be analysed? Thinking back to Experiment One of this chapter and to the precision experiment of Chapter 10 we conclude that a *sampling model* would be more appropriate. The three laboratories can reasonably be regarded as a random sample from all the laboratories likely to use this method. Furthermore, the three batches selected for inclusion of the samples can be regarded as a random sample from a population of batches. Following the approach used in the previous chapter we could draw up a two-way table, as in Table 11.6.

Having tabulated the data we could carry out a two-way analysis of variance. If we follow the procedure that we have used three times already we would have:

(a) 2 degrees of freedom for 'between-laboratories'
(b) 2 degrees of freedom for 'between-batches'

Table 11.6 A tabulation of the data from Experiment Three

Batch	Laboratory					
	A		B		C	
1	d_1	d_2	d_3	d_4	d_5	d_6
2	d_7	d_8	d_9	d_{10}	d_{11}	d_{12}
3	d_{13}	d_{14}	d_{15}	d_{16}	d_{17}	d_{18}

(c) 4 degrees of freedom for 'interaction'
(d) 9 degrees of freedom for 'residual'.

Unfortunately these degrees of freedom would *not* be correct. In fact, the breakdown of the sums of squares to which we have grown accustomed is not appropriate in this situation. To see where we are going astray let us return to the data tabulation, Table 11.6. There is nothing wrong with this table, provided we do not read into it features which do not exist. We must not speak of the three batches in the same way that we spoke of the three physical forms in Table 11.1, or the four levels in the previous chapter. Batch 1 is simply the first batch to be run in each of the three laboratories. We are not speaking of the *same* batch when we speak of batch 1 in lab A and batch 1 in lab B, for example. Perhaps this point will be further clarified if we present the data in a different format (Table 11.7).

Table 11.7 An alternative tabulation for Experiment Three

	Laboratory								
	A			B			C		
Batch	1	2	3	4	5	6	7	8	9
Determination	d_1 d_2	d_3 d_4	d_5 d_6	d_7 d_8	d_9 d_{10}	d_{11} d_{12}	d_{13} d_{14}	d_{15} d_{16}	d_{17} d_{18}

Table 11.7 emphasizes the 'hierarchical or *nested* structure of this experiment compared with the *cross–classified* structure of the previous experiments. We can see in Table 11.4 that to speak of 'between–batches' having two degrees of freedom, is not correct. It would be far more useful to use one-way analysis of variance to break down the variation as in Table 11.8.

'Between-batches' has eight degrees of freedom in Table 11.8 because we have nine batches, not three batches. Table 11.6 is obviously misleading in this respect. Using the mean squares from a completed version of Table 11.8 we could see if the between-batches variation was significantly greater than the within–batch variation. If this was shown to be the case then we would need to bear in mind that

Table 11.8   One-way analysis of variance in
Experiment Three

Source of variation	Degrees of freedom
Between-batches	8
Within-batches	9
Total	17

part of the between-batches variation could be due to having used more than one laboratory. Obviously it would be desirable to separate the between-laboratories variation from the batch-to-batch variation within laboratories. This is achieved in Table 11.9.

Table 11.9   Two-way analysis of variance in Experiment Three

Source of variation	Degrees of freedom
Between-laboratories	2
Between-batches, within-laboratories	6
Residual	9
Total	17

The 'between-batches, within-laboratories' entry arises because of the *nested* nature of the experiment. In each laboratory we have inserted samples into three separate batches, therefore, we have $(3-1=2)$ degrees of freedom for 'between-batches'. This is repeated in each of the three laboratories, giving six degrees of freedom for the 'between-batches, within-laboratories' sum of squares. The other degrees of freedom in Table 11.9 are self-explanatory.

Because we are using a sampling model we must carry out F-tests in the appropriate order. First we test the 'between-batches, within-laboratories' mean-square against the residual mean square. If this is found to be significant we conclude that, within any laboratory, we can expect batch-to-batch variation in excess of the within-batch variation. Thus two determinations on identical samples could be expected to differ more if made in sepatate batches than if made in the same batch. Having found statistical significance at the first F-test we must then compare the 'between-laboratories' mean-square against the 'between-batches, within-laboratories' mean square. If this, in turn, proves to be significant then we conclude that two determinations made in different laboratories are likely to differ more than two determinations made in the same laboratory, even if they were included in different batches.

From the mean squares in the two-way analysis of variance table we could estimate three useful variances. These are:

(a) The testing variance (σ^2) which quantifies the variability within determinations made on identical samples tested in the same batch in the same laboratory.
(b) The between-batches variance (σ_a^2) which quantifies the additional variability introduced by using different batches.
(c) The between-laboratories variance (σ_b^2) which quantifies the additional variability introduced by using different laboratories.

Table 11.10   Interpretations of the model $d = \mu + \beta + \alpha + \delta$

Term in the model	No. of values within the experiment	No. of values within the population	Interpretation
μ	1	1	The true mean determination by this test method when all samples are included in average batches in average laboratories.
β	3	60 but assumed to be ∞	The deviation of a particular laboratory from the average laboratory due to a consistent bias. The variance of all β values would be equal to the between laboratories variance (σ_b^2).
α	9	∞	Deviation of a particular batch from the average batch ($\mu + \beta$) for that particular laboratory. The variance of all α values would be equal to the between-batches variance (σ_a^2).
δ	18	∞	Deviation of a particular determination from the true mean for that particular batch in that particular laboratory. The variance of all δ values would be equal to the testing variance (σ^2).

11.5 Experiment Four

The precision experiment that we discussed in Chapter 10 could be described as a 'uniform-level' experiment. For any one of the sixteen laboratory–level combinations (or cells) the five test samples had the same level of concentration. If there were no random errors introduced by the test method we would, therefore, expect a laboratory to return five identical determinations. Even though there *is* testing error an operator might feel that he *should* produce five identical results if he suspects that the five samples are identical. He might, for example, average the five determinations and enter the mean five times on his results sheet. More subtly, he might increase the lowest determination and decrease the highest, thus reducing the within-cell variance, which would give a reduced repeatability estimate.

British Standard 5497 offers guidance on the conduct of inter-laboratory precision experiments and it offers an alternative type of experiment in which the operator may not be so tempted to adjust his results. It is suggested that if we feel it is unwise to ask an operator to analyse two or more *identical* samples, we might care to set aside the uniform level experiment in favour of the '*split-level*' experiment. In the latter we present the operator with two or more samples that have *different* concentrations; but only *slightly different*.

Let us suppose that the precision experiment from Chapter 10 had been carried out as a split level experiment, using four laboratories and four levels as before, but with each level split into two sub-levels. The experimental design is illustrated in Table 11.11.

Table 11.11 A split-level experiment

		Laboratory			
	Level	A	B	C	D
1a	(9.9%)	d_1	d_3	d_5	d_7
1b	(10.1%)	d_2	d_4	d_6	d_8
2a	(15.1%)	d_9	d_{11}	d_{13}	d_{15}
2b	(15.3%)	d_{10}	d_{12}	d_{14}	d_{16}
3a	(20.0%)	d_{17}	d_{19}	d_{21}	d_{23}
3b	(20.2%)	d_{18}	d_{20}	d_{22}	d_{24}
4a	(24.7%)	d_{25}	d_{27}	d_{29}	d_{31}
4b	(24.9%)	d_{26}	d_{28}	d_{30}	d_{32}

Rather than send, to each laboratory, two identical samples with 10% concentration of ammonia solution, we would send one sample with 9.9% and a second sample with 10.1%. This change in the experimental design will, of course, produce different results. Having used two sub-levels at each of the four main levels we would be very surprised to find two equal determinations in any of the sixteen cells of Table 11.11. The within-cell variation is *partly* due to testing error and *partly* due to the difference between sub-levels. The between-cells variation, on the other hand, is partly due to differences between levels and partly due to differences between laboratories. Thus we have identified four sources of variation without mentioning the word 'interaction'. Obviously the analysis of variance could be quite complicated.

Table 11.12 One-way analysis of variance in
Experiment Four

Source of variation	Degrees of freedom
Between-cells	15
Within-cells	16
Total	31

With a complex situation it is often helpful to start with a one-way analysis of variance. Having separated the between-cells variation and the within-cells variation in Table 11.12 we can then subdivide these two components. The between-cells sum of squares can be split into three parts:

(a) Between-laboratories, with 3 degrees of freedom.
(b) Between-levels, with 3 degrees of freedom.
(c) Interaction, with 9 degrees of freedom.

It should be clear that, as far as laboratories and levels are concerned we have a 'cross-classified sampling model'. We have four laboratories and four levels, with each of the $(4 \times 4 = 16)$ combinations present in Table 11.11. When we come to consider the sub-levels, however, it may not be immediately obvious to the reader whether sub-levels are crossed with levels or nested within levels. We notice that the difference in concentration between the two sub-levels is the *same* at each level (e.g. $10.1 - 9.9 = 0.2\%$, $15.3 - 15.1 = 0.2\%$, etc.). Thus we can regard sub-level as a variable having two values, $+0.1\%$ and -0.1%, and it would therefore be valid to carry out a three-way analysis of variance using a cross-classified, sampling model. This would result in the analysis-of-variance table in Table 11.13.

Table 11.13 A three-way analysis of variance

Source of variation	Degrees of freedom
A. Between-laboratories	3
B. Between-levels	3
C. Interaction (labs × levels)	9
D. Between sub-levels	1
E. Interaction (levels × sub-levels)	3
F. Interaction (labs × sub-levels)	3
G. Interaction (labs × levels × sub-levels)	9
Total	31

In Table 11.13 we have three independent variables: laboratories, levels and sub-levels. Any two of these can be paired to give a two-variable interaction (e.g. labs × levels), and the combination of all three gives a three-variable interaction (labs × levels × sub-levels). The degrees of freedom for the interactions are obtained by multiplying the degrees of freedom of the separate variables. Comparing Table 11.13 with Table 11.12 we see that entries A, B and C in Table 11.13 can be added to give the between-cells entry in Table 11.12, whilst entries D to G can be added to give the within-cells entry.

Note that Table 11.13 does *not* contain a residual. This is only to be expected as no repeat determinations were made in any of the four laboratories. Without a residual how are we to estimate the repeatability of the test method? We will return to this question when we have considered an alternative model.

We justified the three-way cross-classified analysis because the *same* sub-levels (±0.1%) were operative at each level. In practice the achievement of equal intervals at each level might be impossible. Suppose for example the eight concentrations were 9.9%, 10.1%, 15.0%, 15.3%, 19.9%, 20.3%, 24.8%, 24.9%, we could not regard this combination as four levels crossed with two sub-levels. We could, however, regard the concentrations as eight sub-levels nested within four levels, as shown in Fig. 11.1.

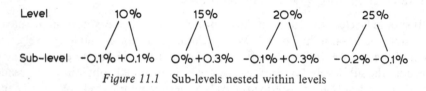

Figure 11.1 Sub-levels nested within levels

Carrying out a three-way analysis of variance, using a model that is partly crossed and partly nested, we obtain the breakdown in Table 11.14.

There are similarities between Table 11.14 and Table 11.13, with the first three entries being identical. The main difference between the two tables lies in the way that the within–cells variation of Table 11.12 has been broken down into component parts. Once again, however, we note the absence of a residual. To

Table 11.14 An alternative three-way analysis

Source of variation	Degrees of freedom
Between-laboratories	3
Between-levels	3
Interaction (labs × levels)	9
Between sub-levels within levels	4
Interaction (sub-levels within levels × labs)	12
Total	31

estimate the repeatability of the test method we must make the assumption that there is *no interaction* between sub-levels and laboratories. It is because of the need to make this assumption that the sub-levels must constitute only *small* deviations (e.g. $\pm 0.1\%$) compared with the differences we can safely assume that a laboratory will perform in a similar way at each of the two sub-levels. If, for example, laboratory A has a certain bias when determining the concentration of a 14.9% solution, we can assume that it will be equally biased with a 15.1% solution.

Having made the assumption that there is no interaction between sub-levels and laboratories we can obtain a residual mean square with twelve degrees of freedom by using either:

(a) The bottom row of Table 11.14, or
(b) A combination of rows F and G in Table 11.13.

The residual mean square will be the same, whichever of the two methods we use, and will form the basis for estimating the repeatability of the test method (see Chapter 10 for details). To estimate the reproducibility of the test method we would also require the 'between-laboratories' mean square and the 'interaction (levels × labs)' mean square. These can be obtained from either Table 11.13 or 11.14.

The reader may have found the above analysis rather heavy-going. It is undoubtedly true that two-way analysis of variance can be much more complex than one-way analysis. For this reason BS 5497 advises us to analyse *each level separately* when considering the results of a precision experiment. (The alternative document IS 4259 strongly recommends the analysis of all levels simultaneously, using two-way analysis, however.)

The eight determinations at the first level in Table 11.11 could easily be used to complete a one-way analysis in which we speak of 'between-laboratories', 'between sub-levels' and 'interaction'. The degrees of freedom would be three, one and three respectively. If we assume that there is no interaction between laboratories and sub-levels then we can refer to the final entry as 'residual' in Table 11.15.

The Residual mean square could be used to estimate repeatability whilst the

Table 11.15   One-way analysis of variance (level 1 only)

Source of variation	Degrees of freedom
Between-laboratories	3
Between sub-levels	1
Residual	3
Total	7

between-laboratories mean square would be needed to estimate the reproducibility of the test method. These estimates would be applicable at the first level only, i.e. 10% concentration. Having obtained similar estimates at all four levels we would need to explore the possibility that the precision of the test method might depend upon the level of concentration being measured.

11.6   Summary

In this chapter we have explored a variety of experiments. In each case we saw how analysis of variance could be used in the analysis of the results of the experiment. Subtle differences between the experimental situations, however, called for modifications to the technique.

We show that some situations require a *sampling* model whilst some require a *complete* model, with occasional need for a *mixed* model when some, but not all, of the variables have values that are sampled. With all three models the sums of squares and the degrees of freedom are obtained by the same formulae but the mean squares are tested against different base lines.

We also distinguished between experimental designs that are *cross-classified* and designs that are *nested*. This is a very important distinction in analytical investigations involving several batches, days, laboratories, operators, reagents, etc.

In the next chapter we will return to the conventional precision experiment that is intended to produce estimates of the repeatability and reproducibility of a test method. We will consider three documents that offer guidance on the conduct of such experiments.

Problems

(1) To evaluate a new analytical method, a precision experiment was carried out in which the same sample was analysed by four operators using both new and old methods. Four replicates were made by each operator with each method.

   (a) Which of the variables require a 'complete model' and which require a 'sampling model'?
   (b) Draw up an analysis of variance table.
   (c) The above experiment was repeated in a second laboratory. With each set of data a sequence of F-tests was carried out, but some of the tests may have been inappropriate. The decisions reached are tabulated below. What conclusions would you draw from each experiment?

Test no.	Mean squares compared	Decisions reached in laboratory A	Decision reached in laboratory B
1	methods v. residual	NS	SIG
2	operators v. residual	SIG	SIG
3	interaction v. residual (methods × operators)	NS	SIG
4	methods v. interaction	NS	NS
5	operators v. interaction	SIG	SIG

NS = Not significant.   SIG = Significant at the 5% level.

(2) A collaborative trial was carried out in seven laboratories over a period of four months on the analysis of hop extract in beer. In the first month, seven cans of beer of type S were selected from the same batch and seven cans from a batch packaged two days previously. One can from each batch was sent to each laboratory which made a single determination on it.

In the second month the sampling procedure was repeated with type L beer. In the third and fourth months respectively, type S and type L beers were analysed using the same sampling scheme.

(a) List the variables.

(b) Which of the variables can be classified as belonging to a complete model?

(c) Which pairs of variables are cross-classified and which are nested?

(d) Draw up an analysis of variance table.

————12————
Precision experiments – advisory publications

12.1 Introduction

This book is an introductory text and does not offer complete answers to all the statistical questions the analyst might ask. Where this book is found to be inadequate the reader should refer to the bibliography for further guidance. This is particularly true for the analyst who contemplates initiating a *precision experiment*.

We have defined a precision experiment as an experiment that is carried out with the purpose of obtaining estimates of the repeatability and reproducibility of a test method. A precision experiment necessarily involves *many* laboratories. A precision experiment that fails to produce *valid* estimates must, therefore, be seen as a large-scale and widespread failure.

In this chapter we will examine three publications which offer detailed guidance on the conduct of precision experiments. They are:

(a) British Standard 5497, Part 1: *Precision of test methods*. (Also published as International Standard ISO/DIS 5725.)
(b) International Standard 4259: *Petroleum products – determination and application of precision data in relation to methods of test*. (Also published as British Standard 4306.)
(c) Youden, W. J. and Steiner, E. H. (1975), *Statistical Manual of the Association of Official Analytical Chemists*.

These three documents will be described in the order in which they are listed above. Throughout the chapter they will be referred to more briefly as:

(a) BS 5497, the British Standard, or simply the standard.
(b) IS 4259, the International Standard, or simply the standard.
(c) Youden and Steiner.

12.2 BS 5497 – organization of the precision experiment

The British Standard offers detailed guidance on the planning, execution, data analysis and reporting of precision experiments. Personnel requirements are clearly stated and the duties of each participant are listed. This is a very useful document and should be studied carefully by anyone embarking on his first inter-laboratory trial. It has been used very successfully in several trials with which the authors have been associated.

The standard recommends that the planning of the precision experiment should be carried out by a *panel* of experts familiar with the test method and its application. This panel should include a statistical expert and an executive officer. Each participating laboratory should nominate a supervisor and an operator. The standard contains a checklist of tasks and problems. This list can very usefully serve as an agenda for panel meetings.

Three fundamental decisions to be made by the panel at the planning stage concern the number of levels (of concentration), the number of replicates and the number of laboratories that will participate in the trial. Obviously it is desirable that these three numbers should be as large as possible but practical constraints will impose limitations. Concerning the number of laboratories, we must not lose sight of the *main* objective of any precision experiment, which is to estimate the *reproducibility* of the test method. The quality of the estimate is closely related to the number of participating laboratories. The British Standard recommends a minimum of eight laboratories if several levels are being investigated and a minimum of fifteen laboratories if only one level is included. (The precision experiments discussed in earlier chapters are grossly deficient in this respect.)

In attempting to obtain the co-operation of an adequate number of laboratories the panel may have to include all laboratories that are prepared to participate. It is highly desirable, however, that the participating laboratories should be representative of *all* laboratories that might use the test method. In the statistical analysis of the data from the experiment it is assumed that we have a random sample of laboratories from an infinite population.

It is not possible to lay down hard and fast rules about the number of levels to be included. In certain circumstances only one level will be needed. If, for example, the test method is to be used solely for checking the conformance of material to a particular specification then all disputes will be centred around the 'minimum acceptable quality' level. It would be reasonable to conduct a precision experiment using just one level and a minimum of fifteen laboratories drawn from supplier and customer laboratories.

Usually, however, a range of levels will be encountered in practice and it is desirable that we include in the experiment a number of levels spread throughout this range. Having included several levels we will wish to ask, at the data-analysis stage, 'Is the precision dependent on the level of concentration?' The need for at least six levels will be very clear when we plot a graph of reproducibility against level of concentration.

Having decided the number of laboratories and the number of levels we can draw up a two-way table, such as Table 9.1. The final decision concerns the number of replicates to include in each cell of the table. The standard recommends that in most situations the minimum number of replicates (i.e. two) will be acceptable. With six levels, eight laboratories and two replicates per cell we have a total of 96 determinations. Increasing the number of replicates to three gives 144 determinations. If resources are available to permit any increase in the number of determinations then it will usually be more beneficial to increase the number of laboratories rather than the number of replicates. Involving a larger number of laboratories will improve the estimate of reproducibility.

12.3   BS 5497 – data analysis

The British Standard advocates a *level-by-level* approach to the analysis of the data from the precision experiment. The data for each level is considered in isolation and estimates of mean determination (m), repeatability (r) and reproducibility (R) are calculated for each level. The precision estimates are then brought together to investigate the possibility that R and/or r might be dependent on m. The whole procedure is illustrated clearly in Fig. 12.1, which is an adaptation of a flow diagram in the British Standard.

The procedure outlined in Fig. 12.1 is equally applicable to the results of a uniform-level experiment or to the results of a split-level experiment, though the method of calculating the precision estimates will differ. The prescribed calculation procedures are equivalent to the use of one-way analysis of variance, though this is not stated in the standard. Before calculation of r and R, however, Cochran's test is applied to the table of standard deviations (or ranges) and Dixon's test is applied to the table of means (treating each level separately, of course).

When we come to investigate the dependence of precision on level of concentration we will obviously wish to plot graphs of R against m, and r against m. The standard suggests that we might explore the relationships by fitting one or more of the following equations:

$$r = a + bm \quad \text{and/or} \quad R = c + dm$$
$$r = bm \quad \text{and/or} \quad R = dm$$
$$\log(r) = a + b \log(m) \quad \text{and/or} \quad \log(R) = c + d \log(m)$$

The statistical expert on the panel might wish to consider other forms of equation but these three have been found useful in practice. The third type of equation is equivalent to $r = Am^b$ or $R = Cm^d$ which offers a variety of curved relationships.

At this point we will reconsider the data from the precision experiment in Chapter 10. The reader may recall that we used two-way analysis of variance to obtain precision estimates then we discussed the assumptions underlying the analysis and used outlier tests to check these assumptions. These tests indicated that the assumptions were satisfied and yet there was a clear indication in Table 10.2 that the precision of the test method was related to the concentration.

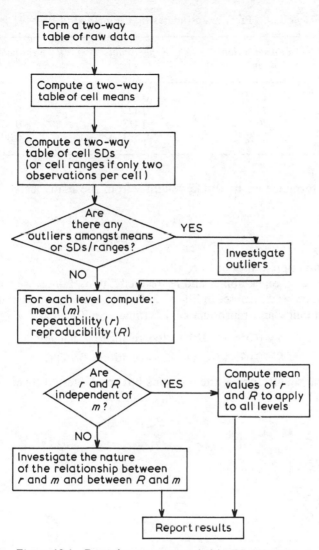

Figure 12.1   Procedure recommended in BS 5497 Part 1

This would seem to violate the assumption that there was a constant 'testing variance', or 'within-cell variance'.

The relationship that we noticed in Table 10.2 will again be revealed if we follow through the prescribed procedure of BS 5497 to obtain the precision estimates in Table 12.1. The precision estimates for level 3 in Table 12.1, do not match exactly those calculated in Chapter 9 ($r = 0.437$ *and* $R = 0.748$). This is

Table 12.1 Precision estimates by the method of BS 5497

Level	Concentration (m)	Repeatability (r)	Reproducibility (R)
1	10%	0.245	0.436
2	15%	0.425	0.522
3	20%	0.413	0.635
4	25%	0.583	0.898

because the formulae we used differ slightly from those advocated in BS 5497 which are:

$$r = 2.83\,\sigma_w$$
$$R = 2.83\sqrt{(\sigma_w^2 + \sigma_b^2)}$$

The 2.83 is an approximation to $t\sqrt{(2)}$.

The estimates of repeatability and reproducibility both increase with increase in concentration, as we can see in Fig. 12.2. Fitting separate regression equations to the repeatability and reproducibility estimates in Table 12.1 we get:

$$r = 0.066 + 0.020m \quad \text{(correlation} = 0.9367)$$
$$R = 0.098 + 0.030m \quad \text{(correlation} = 0.9638)$$

The statistical significance of the relationships can be tested by comparing the

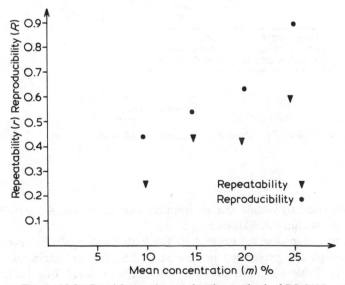

Figure 12.2 Precision estimates by the method of BS 5497

correlation coefficients with critical values from Table D for a sample size of four and a one-sided test. These are:

0.900 at the 5% significance level
0.980 at the 1% significance level

We can conclude that both the repeatability and the reproducibility increase as concentration increases. Having reached this conclusion we would need to act upon it at the reporting stage. When reporting precision estimates for the test method we could present them as in Table 12.1 or give the fitted equations.

To fit the equations relating R and r to m we used *ordinary* regression. There are good reasons, however, for using *weighted* regression in this situation. If, as we appear to have established, the repeatability and reproducibility increase with concentration then the points to the right in Fig. 12.2 are likely to contain larger errors than the points to the left. (It can be shown that the error we are likely to incur whilst estimating a population standard deviation is proportional to the standard deviation that we are attempting to estimate.) If we suspect that the precision of the test method is dependent on the concentration we should use weighted regression with weights related to the true repeatability and reproducibility. This is a chicken-and-egg situation for which the British Standard recommends the use of 'iterative weighted regression'. Our simulations in Chapter 8 would suggest, however, that the use of weighted regression would make little difference to the slope of the fitted line. This proves to be the case with the fitted equations by the iterative method being $r = 0.048 + 0.021m$ and $R = 0.089 + 0.0301m$. We can see in Table 12.2 that the precision estimates are very similar whichever method we use to fit the straight line.

Table 12.2   Final precision estimates by BS 5497

Concentration	Using simple regression		Using weighted iterative regression	
	r	R	r	R
10%	0.266	0.398	0.258	0.390
15%	0.366	0.548	0.363	0.541
20%	0.466	0.698	0.468	0.691
25%	0.566	0.848	0.573	0.842

At this point we will turn our attention to IS 4259 which also offers advice on the design and analysis of precision experiments.

12.4   BS 5497 and IS 4259 – points of difference

Clearly the aim of IS 4259 matches exactly the aim of BS 5497, and it is very likely that many analysts will wish to refer to *both* documents. Unfortunately it is rather difficult to jump from one to the other. This difficulty arises from:

(a) A fundamental difference between the two recommended methods of analysis.
(b) Differences in terminology.
(c) Unstated assumptions in both documents.

Neither document makes reference to the other. Since both were published in 1979, it is reasonable to assume that the two standards were prepared in isolation and it should not be surprising therefore that there are differences in terminology. Perhaps these differences reflect the real difference of interests between the two groups of 'technologists' for whom the documents were produced. Nonetheless, the difficulty of switching from one standard to the other could be greatly reduced by the changing of one or two words in either.

As we have already described British Standard 5497, we will continue to use *its* terminology as far as possible whilst describing International Standard 4259. (This should not be taken to imply that the latter is wrong and the former is correct.) Perhaps this translation will simply postpone an inevitable clash but it will certainly facilitate the flow of the text. Before you attempt to read the International Standard please note that the word *level* is not used therein; the word *sample* being preferred. We will however, continue to use 'level' in the remainder of this description.

Before we examine the recommended method of analysis in IS 4259 we will discuss the reason why one might reasonably adopt a method which differs from that put forward in the British Standard. The fundamental difference between the two standards is actually referred to in the Foreword to BS 5497:

'The method of statistical analysis recommended in section four of this standard is the so-called 'level-by-level' approach rather than the 'overall approach' (which considers an interaction term in the model). The level-by-level method involves carrying out the computations for the mean level of the material, the repeatability and reproducibility for each level separately.'

It appears then that the authors of the British Standard made a conscious decision to adopt the 'level-by-level method'. One advantage of having done so is that the method of analysis is more straight forward. The authors of the International Standard, on the other hand, chose to use the 'overall method'. They *could* claim that complexity of analysis is no disadvantage since computer programs are available to carry out this chore and they could further claim that the use of the overall method gives better estimates of the precision of the test method. One would hope that both authors took into consideration not only the quality of the estimates and the difficulties of computation, but also the currently-accepted practices in their respective fields of influence.

One important difference between the two standards, and a possible source of misunderstanding for anyone reading both documents, lies in the *variances* that are referred to in the methods of data analysis. BS 5497 makes use of:

(a) Within-laboratory variance (σ_w^2).
(b) Between-laboratories variance (σ_L^2).
(c) Repeatability variance (σ_r^2) which is defined as the mean of the within-laboratory variances.
(d) Reproducibility variance (σ_R^2) which is equal to $\sigma_L^2 + \sigma_r^2$.
(e) Repeatability $(r) = 2.83\sigma_r$.
(f) Reproducibility $(R) = 2.38\sigma_R$.

The British Standard attempts to unify these variances by reference to a model, which is written as $y = m + B + e$. Many readers may find this model useful but it must be said that the standard does *not* stick rigidly to the convention of using Greek letters for the population and English letters for the sample. Departing from this convention, whilst using a mixture of Greek and English letters, certainly does *not* help the reader.

The International Standard refers to a rather different set of 'variances' and formulae:

(a) Laboratories standard deviation (D).
(b) Repeats standard deviation (d).
(c) Repeats variance (σ_0^2). (σ_0^2 appears frequently but is not given a name. Repeats variance seems appropriate.)
(d) Interaction variance (σ_1^2).
(e) Between-laboratories variance (σ_2^2).
(f) Repeatability variance $= 2\sigma_0^2$.
(g) Reproducibility variance $= 2(\sigma_0^2 + \sigma_1^2 + \sigma_2^2)$.
(h) Repeatability $= (t\text{-value}) \times (\text{repeatability SD})$.
(i) Reproducibility $= (t\text{-value}) \times (\text{reproducibility SD})$.

Obviously there is ample scope for confusion if one attempts to follow the recommendations of *both* standards. The different definitions of 'repeatability variance' and 'reproducibility variance' in the two documents is particularly unfortunate. Further differences arise when we compare the methods of data analysis, with BS 5497 using one-way analysis of variance at each level separately, and IS 4259 using two-way analysis of variance.

12.5   IS 4259 – data analysis

We introduced two-way analysis of variance in Chapter 10. In fact we applied two-way analysis of variance to the results of a precision experiment given in Table 10.1. Working through the many steps of the procedure we obtained the analysis-of-variance table (Table 10.8). The mean squares in this table were used to estimate three variances (σ_w^2, σ_l^2 and σ_b^2) which in turn were used to obtain precision estimates. The repeatability estimate was equal to 0.433 and the reproducibility estimate equal to 0.647.

Throughout Chapter 10 we were following the recommendations of IS 4259 but for the omission of *one* step, to which we will refer later. The neglect of this one step has resulted in a single estimate being obtained, whereas a range of estimates was obtained when we followed the recommendations of BS 5497.

Table 12.3   Comparison of precision estimates

Concentration	Correct application of BS 5497		Incomplete application of IS 4259	
	r	*R*	*r*	*R*
10%	0.266	0.398	0.433	0.647
15%	0.366	0.548	0.433	0.647
20%	0.466	0.698	0.433	0.647
25%	0.566	0.848	0.433	0.647

When the precision of the test method is related to the concentration and we apply two-way analysis of variance as we did in Chapter 10, then we will get unsatisfactory estimates like those in the right hand columns in Table 12.3. In fairness to IS 4259 we will follow the *complete* method of analysis as set out in Fig. 12.3.

The important step that we missed out in Chapter 10 is in box B of Fig. 12.3. We did *not* transform the data before carrying out the two-way analysis of variance. Whereas the British Standard recommends that we explore the relationship between precision and level of concentration *after* we have calculated repeatability and reproducibility estimates, the International Standard advises that this relationship should be explored *before* the estimates are calculated.

The calculations specified in Fig. 12.3 are more involved and more tedious than those specified in Fig. 12.1. In the first box of Fig. 12.3 we are advised to calculate a between-laboratories standard deviation and a within-laboratory standard deviation for each level. Formulae for this purpose are given in IS 4259. They do not at first sight resemble any formulae given in this text but it can be shown that in effect the author of the International Standard is advising us to carry out a one-way analysis of variance *at each* level. Pursuing this analysis with the data from Table 10.1, we obtain the results shown in Table 12.4.

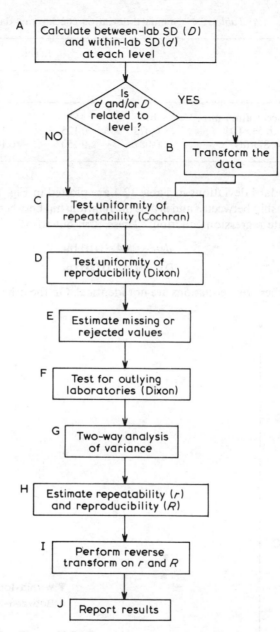

Figure 12.3 Procedure recommended in IS 4259

Table 12.4   Standard deviation check – raw data

	Level			
	1	*2*	*3*	*4*
Mean concentration (m)	10%	15%	20%	25%
Between-lab SD (D)	0.072	0.125	0.171	0.241
Within-lab SD (d)	0.087	0.150	0.146	0.206

The standard deviations in Table 12.4 are plotted in Fig. 12.4. To investigate the relationship between d and m and the relationship between D and m we can fit two separate regression equations. These turn out to be:

$$d = -0.04 + 0.011m$$
$$D = 0.024 + 0.0071m$$

Clearly these two equations are not identical. On the other hand they are not

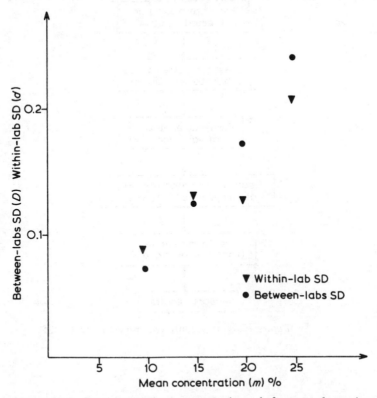

Figure 12.4   Precision against concentration – before transformation

very different. If we were to put forward the hypothesis that the *true* relationships were:

$$d = 0.009m$$

and

$$D = 0.009m$$

you would be unable to reject this hypothesis on the rather limited evidence in Fig. 12.4. (Once again we feel that having six levels rather than four would put us in a much stronger position.)

There is a big advantage to be gained in assuming that the two relationships have identical equations, because if this is the case we need perform only one transformation of the data and not two separate transformations. Let us assume then that $D = 0.009m$ and $d = 0.009m$ are reasonable approximations to the true relationships. As you may already know, we use the logarithmic transformation to obtain homoscedasticity when we find that the standard deviation is proportional to the mean.

If you did not realize that the log transformation was appropriate you could have followed the advice of the International Standard which recommends that when $D = f(m)$ we should transform the data so that:

$$y = K \int \frac{dx}{f(x)} \quad \text{(where } K \text{ is a constant)}$$

Since we have assumed that $D = 0.009m$ then:

$$y = K \int \frac{dx}{0.009x}$$

$$y = \frac{K}{0.009} (\log_e x)$$

or

$$y = \log_{10} x \quad \text{(if } K \text{ is suitably chosen)}$$

The International Standard discusses transformations in some detail including a worked example and a table of useful transformations. Taking logs to the base 10 of the data in Table 12.1 we obtain the transformed data in Table 12.5.

Has the transformation been effective? Let us recall that the transformation was carried out because:

(a) Within-cell variability appeared to increase as the level of concentration increased.
(b) Between-laboratories variation appeared to increase as the level increased.

Inspection of the cell standard deviations in Table 12.6 confirms that the within-cell standard deviations are no longer dependent on level.

The between laboratory variation is determined by the scatter of sample means. Does the standard deviation of cell means vary from level to level? By the

Table 12.5   Transformed data from the precision experiment

	Laboratory			
Level	*A*	*B*	*C*	*D*
1 (10%)	1.000 1.000 1.004 0.996 1.000	0.996 1.000 1.000 1.004 1.000	1.004 1.013 1.000 1.000 1.004	0.996 1.000 0.991 0.996 0.996
2 (15%)	1.182 1.173 1.170 1.179 1.176	1.176 1.185 1.185 1.173 1.176	1.173 1.179 1.179 1.182 1.182	1.170 1.176 1.167 1.170 1.167
3 (20%)	1.294 1.303 1.294 1.294 1.297	1.305 1.307 1.305 1.301 1.307	1.307 1.301 1.299 1.307 1.301	1.299 1.297 1.299 1.299 1.301
4 (25%)	1.394 1.400 1.401 1.394 1.400	1.393 1.400 1.389 1.396 1.394	1.408 1.403 1.400 1.408 1.405	1.405 1.398 1.405 1.401 1.398

Table 12.6   Cell means and standard deviations for the transformed data

	Laboratory			
Level	*A*	*B*	*C*	*D*
1 (10%)	1.000 0 0.002 83	1.000 0 0.002 83	1.004 2 0.005 31	0.995 8 0.003 19
2 (15%)	1.176 0 0.004 74	1.179 0 0.005 61	1.179 0 0.003 67	1.170 0 0.003 67
3 (20%)	1.296 4 0.003 91	1.305 0 0.002 45	1.303 0 0.003 74	1.299 0 0.001 41
4 (25%)	1.397 8 0.003 49	1.394 4 0.004 04	1.404 8 0.003 42	1.401 4 0.003 51

time we have calculated the standard deviation of cell means at each level we may as well complete the one-way analysis of variance four times to get the results in Table 12.7.

Plotting the eight points specified by Table 12.7 gives us Fig. 12.5 which provides the acid test of the effectiveness of our transformation. It certainly appears that we have destroyed the relationship between d and m. We can be confident that the within-laboratory standard deviation is no longer related to

Table 12.7   Standard deviation check – transformed data

	Level			
	1	*2*	*3*	*4*
Mean log (concentration)	1.000	1.176	1.301	1.400
Between-labs SD (*D*)	0.003 01	0.003 74	0.003 63	0.004 19
Within-labs SD (*d*)	0.003 69	0.004 49	0.003 05	0.003 62

the level of concentration. Perhaps we have been less successful with the between-samples standard deviation. Figure 12.5 does indicate a linear relationship between *D* and *m*, though it is not as striking as that in Fig. 12.4.

A significance test reveals that the relationship between *D* and *m* in Fig. 12.5 is not statistically significant. We could conclude that the transformation has been effective or we could decide to employ a different transformation. We must again feel that four points is inadequate for our purpose and resolve never again to carry out a precision experiment with four levels.

If we conclude that the transformation has been effective then the next step in

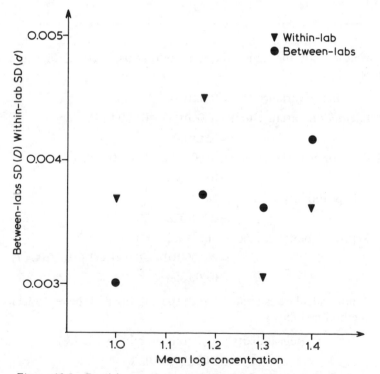

Figure 12.5   Precision against concentration – after transformation

the analysis is to carry out the tests in boxes C, D and F of Fig. 12.3. In the International Standard these activities are listed under the heading *Tests for Outliers*. Full details will not be given as we performed similar tests in Chapter 10, but no outliers are found with this set of data.

Having checked many of the assumptions which underlie two-way analysis of variance we can now get down to using this technique to obtain precision estimates from our transformed data in Table 12.5. (If our outlier tests had yielded many rejections we would first need to return to the raw data to check the appropriateness of our chosen transformation.) The complete analysis of variance table is shown in Table 12.8.

Table 12.8   Two-way analysis of variance on transformed data

Source of variation	Sum of squares	Degrees of freedom	Mean square
Between-levels	1.783 347	3	0.594 449
Between-labs	0.000 456	3	0.000 152
Interaction	0.000 521	9	0.000 059
Within-cells	0.000 888	64	0.000 014
Total	1.785 212	79	

Following exactly the same procedure as that used earlier we can estimate variances as follows:

$$\text{Error variance } (\sigma_e^2) = 0.000\,014$$
$$\text{Interaction variance } (\sigma_I^2) = (0.000\,059 - 0.000\,014)/5$$
$$= 0.000\,009$$
$$\text{Laboratories variance } (\sigma_B^2) = (0.000\,152 - 0.000\,059)/20$$
$$= 0.000\,004\,7$$
$$\text{Repeatability variance} = 2\sigma_e^2$$
$$= 0.000\,028$$
$$\text{Reproducibility variance} = 2(\sigma_e^2 + \sigma_B^2 + \sigma_I^2)$$
$$= 2(0.000\,014 + 0.000\,009 + 0.000\,004\,7)$$
$$= 0.000\,055\,4$$

We can now calculate estimates of the repeatability and the reproducibility of the test method as follows:

$$\text{Repeatability} = t_{64}\sqrt{(\text{repeatability variance})}$$
$$= 2.00\sqrt{(0.000\,028)}$$
$$= 0.0106$$

$$\text{Reproducibility} = t_{30}\sqrt{(\text{reproducibility variance})}$$
$$= 2.04\sqrt{(0.000\,055\,4}$$
$$= 0.0152$$

These precision estimates are applicable to the transformed data (y). To obtain estimates which are applicable to the raw data (x) we must introduce the reverse transformation:

$$(\text{repeatability of } x) = \left|\frac{dx}{dy}\right| (\text{repeatability of } y)$$

Since our transformation was $y = \log_{10} x$ which can be written $y = 0.4343 \log_e x$, then:

$$\frac{dy}{dx} = 0.4343\left(\frac{1}{x}\right)$$

and

$$\frac{dx}{dy} = \frac{x}{0.4343}$$

Hence:

$$(\text{repeatability of } x) = \left(\frac{x}{0.4343}\right) (\text{repeatability of } y)$$
$$= (2.303x)(0.0106)$$
$$= 0.0244x$$

To find the repeatability of our test method at any particular mean level (x) we simply multiply the level by 0.0244. This has been done to obtain the estimates tabulated in Table 12.9.

Table 12.9   Repeatability and reproducibility of the test method

	Level of concentration x			
	10%	*15%*	*20%*	*25%*
Repeatability = 0.0244x	0.244	0.366	0.488	0.610
Reproducibility = 0.0350x	0.350	0.525	0.700	0.875

The above estimates in Table 12.9 which have been obtained by following the procedure recommended in IS 4259, should be compared with the estimates in Table 12.3. The latter were obtained by following the rather different procedure recommended in BS 5497. Clearly there is little difference between the two sets of estimates despite the very different routes that we have taken to reach them. Once again, perhaps, we feel that estimates of repeatability and reproducibility at *more than four* levels would provide a better basis for comparing the two methods of analysis. You will recall that the British Standard recommended the use of at least six levels in a precision experiment. We will now explore the recommendations of the International Standard with respect to the design of precision experiments.

IS 4259 – design of the precision experiment

In any precision experiment we will have several laboratories testing material at several levels with two or more repeats at each level. In designing the experiment we must decide how many laboratories, levels and repeats we are going to use.

Throughout the International Standard it is assumed that in *every* precision experiment *only two repeats* will be used at each laboratory level combination. This assumption is not clearly stated, but it is there nonetheless. Is it a reasonable assumption? Certainly it is a very common practice to use the minimum number of repeats (i.e. two) especially if the analytical procedure is time consuming or costly. On the other hand there do exist situations in which the measurement procedure is almost trivial and where more than two repeats would seem desirable.

When designing a precision experiment we must not lose sight of the fact that the end product is a set of precision estimates. It is highly desirable that both the repeatability estimates and the reproducibility estimates should be based on a large number of degrees of freedom. The International Standard recommends that we aim for at least *thirty* degrees of freedom for each. With only two repeats per cell we get one degree of freedom within each cell and the repeatability estimate will have the necessary degrees of freedom provided we use at least thirty cells (i.e. thirty laboratory-level combinations).

If we decided to use thirty cells, would it be better to use six levels and five laboratories or ten levels and three laboratories or invoke some other pair of numbers which could be multiplied to give a product of thirty? The International Standard offers some very useful advice on this point. In fact IS 4259 contains a table which is specifically designed to help us decide the number of levels that are needed for any particular number of laboratories if we wish to have at least thirty degrees of freedom for our reproducibility estimates. In order to make use of this table, however, we need estimates of the between-laboratories variance and the interaction variance.

Such estimates will not, of course, be available in most cases. The International Standard suggests that we carry out a *pilot experiment* in order to obtain the variance estimates. Our first problem then is how to design the pilot experiment. The standard recommends the use of at least two levels, at least twelve laboratory-level combinations and, of course, two repeats in each cell.

A pilot experiment will not only yield the variance estimates needed to plan the main precision experiment, it will also enable us to check:

(a) Operators' instructions for the test method.
(b) Bias of the test method.
(c) Logistics, safety precautions, etc.

If the pilot experiment revealed excessive bias or unexpectedly low precision we might, of course, decide to modify the method of test rather than proceed to the main experiment.

Assuming that the pilot programme confirms the validity of the test method we can calculate estimates of the between-laboratories variance and the interaction variance, then we can use Table 11 in IS 4259 to help us reach a decision about the number of laboratories and the number of levels to be used in the precision experiment.

12.7   Youden and Steiner

This small book, though published as *one* text by the AOAC, consists of an explanatory foreword followed by two distinctly different 'papers':

(a) *Statistical techniques for collaborative tests*, by W. J. Youden (63 pages).
(b) *Planning and analysis of results of collaborative tests*, by E. H. Steiner (20 pages).

Youden, who trained as a chemist but later became a statistician of some eminence, offers advice that every scientist can understand and should study carefully before embarking on a precision experiment. Youden's vast experience in this area is crystallized into this small volume and many of the points he makes complement the recommendations of the two documents discussed earlier in this chapter.

Youden concentrates upon the *split-level experiment*, though he does not refer to it as such. When we introduced the split-level experiment in Chapter 11 using two-way analysis of variance, the reader may have found the discussion rather tedious. Our main purpose at that time was to focus on the assumptions underlying the technique. The calculation of precision estimates from the results of a split-level experiment can, however, be very simple if we follow the procedure recommended in BS 5497 or in Youden and Steiner.

The essential step in the simplification is to tabulate the results of the experiment as in Table 11.11, then to calculate for each cell the *total* of the two results and the *difference* between the two results. If the errors in the determinations are purely random then we would expect the cell differences to be just as variable as the cell totals. If, on the other hand, systematic errors are introduced by one or more laboratories, then the cell differences will be *less* variable than the cell totals, as Youden emphasizes. By calculating the standard deviation of cell totals and the standard deviation of cell differences we can obtain estimates of the repeatability and reproducibility of the test method.

Youden recommends, and we would wholeheartedly agree, that the results of the precision experiment be graphed for visual inspection *before* any calculations are carried out. A simple scatter diagram, or 'two-sample chart' should be drawn for each level. If the scatter of the test results is due solely to random errors we would expect a scatter as in Fig. 12.6(a). In Youden's experience with many

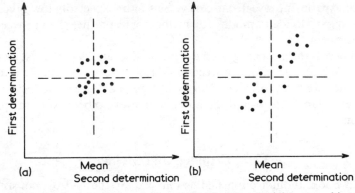

Figure 12.6   A 'two-sample chart'. (a) No bias, (b) laboratory bias

hundreds of scatter diagrams, however, he hasn't found a single instance of such an equal spread of points (i.e. laboratories) between the four quadrants. Invariably the points fall mainly into two quadrants because of laboratory bias, to give a scatter as in Fig. 12.6(b).

An extremely important point made by Youden, which is *not* mentioned in BS 5497 or IS 4259, concerns the need to investigate the 'ruggedness' of the test method *before* committing several laboratories to the considerable cost and inconvenience of an inter-laboratory trial. In Youden's experience '. . . the results from many collaborative tests are extremely disappointing. After a lot of excellent work has been done, it turns out that the procedure at that stage of its development has serious flaws, or perhaps the details of the various steps are inadequately specified. The result is a totally unacceptable lack of reproducibility among the results of different laboratories.'

Such unfortunate results are likely to be avoided if the initiating laboratory is prepared to carry out a 'ruggedness experiment' before declaring the test method to be ready for an inter-laboratory trial. In this experiment small variations in the test method are deliberately introduced. These changes should be of a magnitude that could be expected to occur (inadvertently) when the method is used in different laboratories. Youden offers details of an experimental design that is particularly suited to this purpose. For a broader discussion of experimental design the reader is referred to Caulcutt (1983) and Davies (1978).

An additional test advocated by Youden, and not included in BS 5497 or IS 4259, is the 'ranking test for laboratories'. At each level, the *n* laboratories are ranked from 1 to *n*, with rank 1 being assigned to the laboratory with the highest determination and rank *n* to the laboratory with the lowest. A total rank is then calculated for each laboratory. Attention is then focused on any laboratory with very high or very low total rank. Use of a table of critical values can be used to decide whether or not these rank totals are more extreme than could be expected by chance considering the number of laboratories and the number of levels.

When using the ranking test or when plotting the 'two-sample chart' (Fig. 12.6) the analyst is again made aware of the need to include as many laboratories as possible in the precision experiment. Youden recommends the inclusion of ten or more laboratories in a collaborative study intended to evaluate the performance of a test method in the 'average laboratory'.

Turning now to the other half of this publication, we see that Steiner's recommendations are rather similar to those of IS 4259 and are applicable to what was earlier described as the 'uniform-level experiment'. He advocates the use of two-way analysis of variance preceded by an exploration of the relationship between precision and level. The two-way analysis allows the analyst to distinguish between 'consistent laboratory differences' and 'random labora-tory differences'; the latter manifesting itself in the form of an interaction between laboratories and levels.

Steiner would disagree with Youden on the wisdom of using a split-level experiment. In fact Steiner goes so far as to say that 'It is impossible to measure the repeatability of a method unless replicate determinations are made in different laboratories'. We saw in Chapter 11 that we could estimate both repeatability and reproducibility from the results of the split-level experiment only if we were prepared to make certain assumptions.

12.8 Summary

In this chapter we have examined three publications that offer detailed guidance on the conduct of inter-laboratory trials. Before even planning such an experiment the analyst would be well advised to consult Youden and Steiner (1975) and either BS 5497 or IS 4259.

─────13─────

Miscellaneous topics

13.1 Introduction

In this chapter we will discuss two topics that could have been introduced much earlier. The first of these, the method of standard additions, would have fitted quite logically into Chapter 7 if we had posed the question 'Do the calibration standards have the same matrix as the operational samples?' The second topic, limit of detection, could have been examined as early as Chapter 3 if we had asked 'With random errors influencing every sample, even blanks, how can we be sure that a small reading is telling us anything at all?'

13.2 The method of standard additions

In our discussion of calibration it was assumed that the calibration standards and the unknown sample were similar in every respect except the concentration of determinand. The analytical chemist knows that this cannot always be realized in practice. It is possible, for example, that the calibration standards are very pure whereas the unknown sample contains an interfering matrix, with the result that a proportional error is introduced. In such situations the method of standard additions can be useful.

When using the method of standard additions the unknown sample is split into several samples which are then spiked with different (and known) quantities of the determinand. Suppose, for example, that we wish to determine the concentration of cuprammonium ion in a sample. Splitting the sample into five samples we add 0.012 M to the first, 0.009 M to the second, 0.006 M to the third, 0.003 M to the fourth and leave the fifth sample undisturbed. The absorbance of each of the five samples is now measured to give the results in Table 13.1.

Table 13.1

Added cuprammonium	0.000	0.003	0.006	0.009	0.012
Absorbance	0.33	0.45	0.61	0.74	0.85

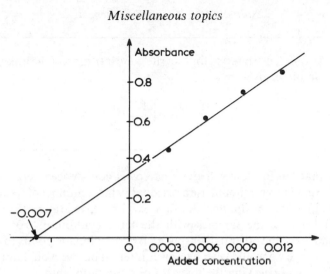

Figure 13.1 Data from the method of standard additions

The data in Table 13.1 is plotted in Fig. 13.1, together with the ordinary least-squares regression line. (Once again we are assuming that the precision of the test method is *not* related to the concentration of the determinand.)

The intercept of the regression line on the *x*-axis indicates the concentration of cuprammonium in the original sample. The figure of -0.007 tells us that we would need to add -0.007 M (i.e. remove 0.007 M) in order to get an absorbance reading of zero. If we wish to obtain this result without plotting the graph we can use the formula:

$$\text{Predicted concentration} = \frac{a}{b}$$

where a is the intercept, and
b is the slope of the regression line

The equation of the regression line is $y = 0.3300 + 44.333x$, giving:

$$\text{Predicted concentration} = \frac{a}{b} = \frac{0.3300}{44.333}$$

$$= 0.007\,443\,6 \text{ M}$$

This prediction will, of course, be in error. The random errors in the five absorbance measurements have been propagated into the calculated slope and intercept and hence into the predicted concentration. We can, however, calculate a confidence interval for the true concentration.

A confidence interval for the true concentration of the unknown sample is given by:

$$\frac{a}{b} \pm \frac{t(\text{ESD})}{b}\sqrt{\left[\frac{1}{n}+\left(\frac{\bar{y}}{b}\right)^2\bigg/ \text{S}xx\right]}$$

We can see that this formula will give a narrower confidence interval if n is large and Sxx is large. Thus we should use a reasonably large number of spiked samples (say $n=6$ to 8). For a maximum value of Sxx, with a given range of additions, we should spike half of the samples with the largest addition (0.012 M in this example) and leave the other half unspiked. By crowding the points to the ends of the working range we would get a narrower interval but we would not, of course, have any means of checking the linearity of the relationship.

Using the residual standard deviation (0.014 49) as an estimate of the error standard deviation (ESD) and substituting $a=0.3300$, $b=44.333$, $t_3 = 3.18$, $n=5$, $\bar{y}=0.596$ and S$xx=0.000\,09$ we get:

$$95\% \text{ confidence interval} = \frac{0.33}{44.33} \pm \frac{3.18(0.014\,49)}{44.333}\sqrt{\left[\frac{1}{5}+\left(\frac{0.596}{44.333}\right)^2\bigg/ 0.000\,09\right]}$$

$$= 0.007\,44 \pm 0.001\,039\sqrt{(0.2+2.0081)}$$

$$= 0.007\,44 \pm 0.001\,54 \text{ M}$$

$$= 0.005\,90 \text{ M} \quad \text{to} \quad 0.008\,98 \text{ M}$$

This confidence interval can usefully be compared with those in Table 7.2. When using the calibration line we obtained a confidence interval of $0.006\,24 \pm 0.002\,27$ M for a sample which had an absorbance reading of 0.30. Obviously this is substantially in agreement with the interval given by the method of standard additions. Furthermore the residual standard deviations (0.027 18 in Chapter 7 and 0.014 49 on this occasion) are rather close whilst the slopes of the regression lines (41.667 in Chapter 7 compared with 44.333) differ very little.

The close agreement between the two slopes implies that the use of the method of standard additions is not necessary with this particular test method. On the other hand, if we had found a large difference between the slopes, this would have suggested that the use of the calibration line was not valid and that the method of standard additions would give more accurate estimates of concentration. Obviously, the extra effort involved with standard additions should not be undertaken without sound evidence that it is necessary, and the difference between the two slopes can be checked by means of a significance test. This proceeds as follows:

Null hypothesis – The true slopes of the two lines are equal ($\beta_1 = \beta_2$).

Alternative hypothesis – The true slopes of the two lines differ ($\beta_1 \neq \beta_2$).

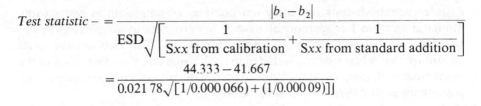

$$\text{Test statistic} - = \frac{|b_1 - b_2|}{\text{ESD}\sqrt{\left[\dfrac{1}{S_{xx} \text{ from calibration}} + \dfrac{1}{S_{xx} \text{ from standard addition}}\right]}}$$

$$= \frac{44.333 - 41.667}{0.021\,78\sqrt{[1/0.000\,066) + (1/0.000\,09)]]}}$$

$$= 0.755.$$

Critical values – from the two-sided t-table using 6 degrees of freedom are:

  2.45 at the 5% significance level
  3.71 at the 1% significance level.

Decision – We are unable to reject the null hypothesis.

Conclusion – We are unable to conclude that the true slopes of the two lines differ.

Note: In calculating the test statistic the error standard deviation (ESD) was estimated by combining the residual standard deviations from the two regression lines:

$$\text{ESD} = \sqrt{\left[\frac{\text{d.f.}(\text{RSD})^2 + \text{d.f.}(\text{RSD})^2}{\text{d.f.} + \text{d.f.}}\right]}$$

$$= \sqrt{\left[\frac{3(0.014\,49)^2 + 3(0.027\,18)^2}{3+3}\right]}$$

$$= 0.021\,78$$

This estimate of error standard deviation has 6 degrees of freedom, a fact which was made use of in obtaining the critical values from the t-table.

A test statistic greater than 2.45 would have indicated that there was a significant difference between the two slopes, and hence a need to abandon the calibration line and adopt the method of standard additions.

13.3 Limit of detection

The limit of detection of a method may be defined as:

> The smallest concentration of a determinand for which we can be 95% confident that the determinand will be detected by the method.

The confidence level of 95% can obviously be altered by the analyst to any level he considers appropriate. This level does however represent a risk of making a wrong decision and should be related to how the determinations are to be used rather than to the accuracy of the method.

Independent Analysts Ltd carry out analyses of potassium in water using atomic absorption. For operational reasons they group many samples together in batches and analyse them over a short time period. This also has the great advantage that a blank can be included in each batch and the subtraction of the blank from each sample reading helps to reduce the bias to a negligible level. The reading for each determination is therefore:

$$\text{Corrected reading} = \text{sample reading} - \text{blank reading}$$

The concentration is found from a calibration equation which in this situation has been estimated to be:

$$\text{Concentration} = 2.54 \ (\text{corrected reading})$$

The calibration equation was obtained from a calibration experiment which confirmed that there was a linear relationship between the variables throughout the range of concentrations in the experiment and that this relationship holds even at very small concentrations. We see that the regression line goes through the origin.

Before computation of the limit of detection it is necessary to define a *criterion of detection* which is the corrected reading for a determination, above which we will be confident of the presence of the substance that we hope to detect. When he obtains a corrected reading below the criterion the analyst will report that he is unable to detect the presence of potassium, whilst a reading above the criterion will be converted into a concentration and reported as such. To obtain the criterion of detection we must determine the standard deviation of blanks which is obtained from duplicate blank readings in several batches, usually spread over several days. These readings may include negatives and must be made with the finest discrimination possible. Independent Analysts use eight batches and obtain the following values:

Table 13.2

Batch	Blanks		Diff	Diff^2
1	0.171	0.031	0.140	0.019 600
2	−0.061	0.020	0.081	0.006 561
3	0.003	0.021	0.018	0.000 324
4	0.141	−0.063	0.204	0.041 616
5	0.072	0.146	0.074	0.005 476
6	0.031	−0.061	0.092	0.008 464
7	0.011	0.042	0.031	0.000 961
8	0.063	0.011	0.052	0.002 704
				0.085 708

Standard deviation of corrected blanks $(s_b) = \sqrt{\left[\dfrac{\sum(\text{diff})^2}{n}\right]}$

$$= \sqrt{(0.085\,706/8)}$$

$$= 0.104$$

The definition of the criterion of detection implies that 5% of samples with zero concentration will give a reading above the criterion. Shown below is a diagrammatic interpretation of the criterion which also shows that it is based on the assumption that errors in readings follow a normal distribution.

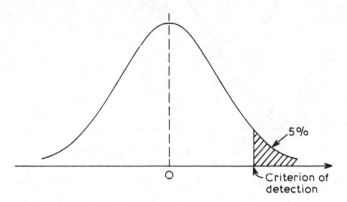

Figure 13.2   Calculation of criterion of detection

Since we have estimated the standard deviation we use the *t*-distribution in the computation with 8 degrees of freedom:

$$t_b = 1.86$$

$$\text{Criterion of detection} = t_b s_b$$

$$= 1.86 \times 0.104$$

$$= 0.193 \text{ microamps}$$

It is now necessary to carry out an experiment to estimate the standard deviation (s_d) of readings made on samples which have a concentration at the limit of detection. We do not know, of course, the limit of detection, so a concentration must be used which is fairly close to it. If subsequently this concentration is found to differ greatly from our new estimate of the limit of detection then the experiment may have to be repeated. A method of estimating the concentration for the experiment is as follows.

If we assume that s_d is equal to s_b then the reading corresponding to the limit of detection will be twice that of the criterion of detection. This gives a value of $0.193 \times 2 = 0.386$. However, s_d is likely to be higher than s_b and it may be prudent to add 20% onto the microamp reading. This gives a microamp reading of 0.463.

Using the calibration equation gives a concentration of approx. 1.20 mg/litre.

A sample at this concentration was included in eight different batches and the corrected readings, found by subtracting the blank reading from the sample reading, were:

0.516   0.407   0.273   0.501   0.391   0.582   0.203   0.367

These give a standard deviation (s_d) of 0.127 based on 7 degrees of freedom.

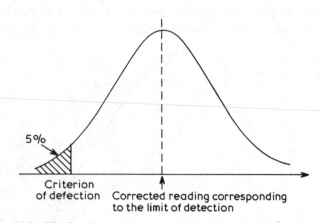

5%

Criterion
of defection   Corrected reading corresponding
to the limit of detection

Figure 13.3   Finding the reading corresponding to the limit of detection

The corrected reading corresponding to the limit of detection (microamps) is computed so that there is only a 5% chance that a sample at the limit of detection will give a reading below the criterion of detection. This is given by:

Corrected reading corresponding to the limit of detection

$$= \text{criterion of detection} + t_d s_d$$
$$= 0.193 + 1.89 \times 0.127$$
$$= 0.433$$

Using the calibration equation:

Limit of detection $= 2.54$ (corrected reading)
$$= 2.54 \times 0.433$$
$$= 1.099\,82$$

We can therefore quote the limit of detection as 1.1 mg/litre.

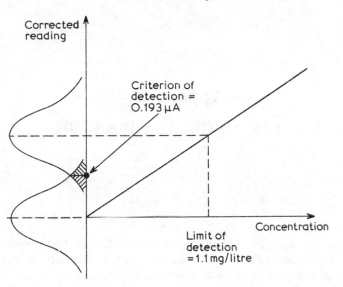

Figure 13.4 Calculating the limit of detection

13.4 Summary

In this chapter we have made use of several statistical techniques to shed light on two problems which are familiar to all analytical chemists. To estimate the limit of detection we quantified the variability in repeat readings made on blanks, and we used values from the t-table with appropriate degrees of freedom. To detect the presence of interference effects we compared the slope of a calibration line with the slope of a line fitted to data obtained by the method of standard additions. In cases where the two slopes are significantly different we could expect the method of standard additions to give more accurate predictions of the concentration of determinand in an unknown sample.

The reader is reminded, at this point, that *all* of the statistical techniques introduced in this book can be used in *other* situations, sometimes very different from those which we used for illustration. Furthermore, we would like to caution the reader that this book should be regarded as an introduction rather than as a handbook and further reading from the texts listed in the bibliography is strongly recommended.

Appendix 1
The sigma (Σ) notation

Throughout this book the calculation of certain statistics involves adding sets of numbers. The simplest case is where we add a set of observations. For example, suppose that the number of books purchased by six analytical chemists during 1983 was 0, 4, 2, 2, 1, 4. Adding these six figures we obtain a total of 13 books. We can either refer to this total as 'the sum of the observations' or we can use a statistical shorthand which is far more concise. Thus $\sum x$ is the shorthand for 'the sum of the observations' where x is the symbol for an observation and $\sum$ is the symbol for *add*. Therefore $\sum x$ is an instruction telling us to 'add the observations'.

$$\sum x = 0 + 4 + 2 + 2 + 1 + 4 = 13$$

We can now use this shorthand to represent other statistics. For example 'the sum of the squared observations' is denoted by $\sum x^2$. Using the data given above:

$$\sum x^2 = 0^2 + 4^2 + 2^2 + 2^2 + 1^2 + 4^2 = 41$$

For another example let us calculate $\sum(x-2)^2$. This gives:

$$\sum(x-2)^2 = (0-2)^2 + (4-2)^2 + (2-2)^2 + (2-2)^2 + (1-2)^2 + (4-2)^2 = 13$$

The gain in simplicity and unambiguity can clearly be seen if we compare the mathematical expression with the written expression which is 'the sum of squares of the observations after two has been subtracted from each observation'.

———Appendix 2———
Nomenclature and formulae

Nomenclature

n	sample size, number of observations or number of points
$\bar{x}$	sample mean
s	sample standard deviation
s^2	sample variance
μ	population mean
σ	population standard deviation
σ^2	population variance
σ_b^2	between-laboratories variance
σ_I^2	interaction variance
σ_w^2	within-laboratory variance
z	standardized value
t	critical value from the t-table
c	a change (or difference) which we wish to detect
$\sum$	'the sum of'
d.f.	degrees of freedom
r	correlation coefficient or repeatability estimate
R	reproducibility estimate
a	intercept $\left.\right\}$ of least-squares regression line, $y = a + bx$
b	slope
RSS	residual sum of squares
RSD	residual standard deviation
ESD	error standard deviation
Sxx	sum of squares
Sxy	sum of cross products
$Swxx$	weighted sum of squares
$Swxy$	weighted sum of cross product
CI	confidence interval

Formulae

Sample mean: $\bar{x} = \sum x / n$

207

Sample standard deviation:

$$s = \sqrt{[\sum(x-\bar{x})^2/(n-1)]} = \sqrt{[(\sum x^2 - n\bar{x}^2)/(n-1)]}$$

Coefficient of variation $= (s/\bar{x}) \times 100\%$

Variability of sums and differences:

$$SD(x \pm y) = \sqrt{[SD(x)^2 + SD(y)^2]} \text{ if } x \text{ and } y \text{ are independent}$$
$$SD(a+x) = SD(x) \quad \text{where } a \text{ is a constant}$$

Variability of products and quotients:

$$CV(x \overset{\times}{\div} y) = \sqrt{[CV(x)^2 + CV(y)^2]}$$
$$SD(ax) = a\,SD(x) \quad \text{where } a \text{ is a constant}$$

Confidence interval for μ: $\bar{x} \pm ts/\sqrt{n}$

Confidence interval for σ: $L_1 s$ to $L_2 s$

Sample size needed to estimate a bias to within $\pm c$:

$$n \simeq (ts/c)^2$$

Sum of squares:

$$Sxx = \sum(x-\bar{x})^2 \quad \text{or} \quad \sum x^2 - n\bar{x}^2 \quad \text{or} \quad (n-1)(SD \text{ of } x)^2$$

Sum of cross products:

$$Sxy = \sum(x-\bar{x})(y-\bar{y}) \quad \text{or} \quad \sum xy - n\bar{x}\bar{y}$$

Correlation coefficient: $r = \dfrac{Sxy}{\sqrt{(Sxx\,Syy)}}$

Percentage fit: $100\,(\text{correlation})^2$

Combined estimate of population standard deviation:

$$\sqrt{\left\{\frac{\sum[(\text{d.f.})(SD)^2]}{\sum\text{d.f.}}\right\}}$$

Least squares regression line, $y = a + bx$

Slope $(b) = Sxy/Sxx$
Intercept $(a) = \bar{y} - b\bar{x}$
Residual sum of squares for a regression line:

$$RSS = Syy - b^2 Sxx$$

Residual standard deviation:

$$RSD = \sqrt{[(\text{residual sum of squares})/(\text{degrees of freedom})]}$$

Confidence interval for the true intercept (α):

$$a \pm t(\text{ESD}) \sqrt{\left(\frac{1}{n} + \frac{\bar{x}^2}{\text{S}xx}\right)}$$

Confidence interval for the true slope (β):

$$b \pm t(\text{ESD}) / \sqrt{\text{S}xx}$$

Confidence interval for the true mean y value corresponding to a specified x value (X):

$$(a + bX) \pm t(\text{ESD}) \sqrt{\left[\frac{1}{n} + \frac{(X - \bar{x})^2}{\text{S}xx}\right]}$$

Confidence interval for an individual y value corresponding to a specified x value (X):

$$(a + bX) \pm t(\text{ESD}) \sqrt{\left[1 + \frac{1}{n} + \frac{(X - \bar{x})^2}{\text{S}xx}\right]}$$

Confidence interval for an individual x value corresponding to the mean of m values of y ($\bar{Y}$):

$$\left(\frac{\bar{Y} - a}{b}\right) \pm \frac{t(\text{ESD})}{b} \sqrt{\left[\frac{1}{m} + \frac{1}{n} + \left(\frac{\bar{Y} - \bar{y}}{b}\right)^2 \Big/ \text{S}xx\right]}$$

Least squares regression line through the origin, $y = bx$

Slope (b) $= \sum xy / \sum x^2$
Residual sum of squares $= \sum y^2 - b^2 \sum x^2$
Confidence interval for the true slope (β):

$$b \pm t(\text{ESD}) / \sqrt{\sum x^2}$$

Confidence interval for the value of x corresponding to a specified value of y (Y):

$$\frac{Y}{b} \pm \frac{t(\text{ESD})}{b} \sqrt{\left[1 + \left(\frac{Y}{b}\right)^2 \Big/ \sum x^2\right]}$$

Weighted least squares regression line, $y = a \pm bx$

Weighted mean $\bar{x}_w = \sum wx / n$
Weighted sum of squares $(\text{S}wxx) = \sum wx^2 - n\bar{x}_w^2$
Weighted sum of cross products $(\text{S}wxy) = \sum wxy - n\bar{x}_w\bar{y}_w$
Slope (b) $= \text{S}wxy / \text{S}wxx$

Intercept $(a) = \bar{y}_w - b\bar{x}_w$

Confidence interval for the value of x corresponding to a specified value of y (Y):

$$\left(\frac{Y-a}{b}\right) \pm \frac{t(\text{ESD})}{b}\sqrt{\left[\frac{1}{w} + \frac{1}{n} + \left(\frac{Y-\bar{y}_w}{b}\right)^2 \middle/ Swxx\right]}$$

Residual standard deviation:

$$\sqrt{[(Swyy - b^2 Swxx)/(n-2)]}$$

Analysis of variance

Total sum of squares $= (\text{SD of all determinations})^2 \times (\text{d.f.})$

Within-lab sum of squares $= \sum[(\text{SD of determinations in one lab})^2 \times (\text{d.f.})]$

Between-labs sum of squares $= (\text{SD of lab means})^2 \times (\text{d.f.}) \times (\text{no. of determinations per lab})$

Repeatability estimate $(r) = 1.96\sqrt{(2)}\sigma$ or $t\sqrt{(2)}s$

Reproducibility estimate $(R) = 1.96\sqrt{(2)}\sqrt{(\sigma_w^2 + \sigma_b^2)}$

or $t\sqrt{(2)}\sqrt{(\text{estimate of } \sigma_w^2) + \text{estimate of } \sigma_b^2)}$

Significance testing

Significance test	Test statistic
One-sample t-test	$\dfrac{\lvert \bar{x} - \mu \rvert}{s/\sqrt{n}}$
t-test for curvature	$\dfrac{\lvert \text{residual from upper point} \rvert}{\text{RSD}\sqrt{\left[\dfrac{(n+1)}{n}\right]}}$
F-test	Larger variance/smaller variance
Dixon's test	The larger of A or B below
Sample size between 3 and 7 inclusive	$A = \dfrac{x_2 - x_1}{x_n - x_1}$ $B = \dfrac{x_n - x_{n-1}}{x_n - x_1}$
Sample size between 8 and 12 inclusive	$A = \dfrac{x_2 - x_1}{x_{n-1} - x_1}$ $B = \dfrac{x_n - x_{n-1}}{x_n - x_2}$
Sample size greater than 12	$A = \dfrac{x_3 - x_1}{x_{n-2} - x_1}$ $B = \dfrac{x_n - x_{n-2}}{x_n - x_3}$
Cochran's test	$\dfrac{\textit{Largest variance}}{\sum \text{variances}}$

─────────Solutions to problems─────────

Chapter 2

Solution to Problem 1

(a)

Sample	A	B	C
Mean	6.20	36.90	243.90
Standard deviation	0.224	1.098	5.686
Coefficient of variation	3.61%	2.97%	2.33%

(b) These results are fairly typical for analytical methods. As the mean increases the lack of precision, as measured by the standard deviations, greatly increases but the relative lack of precision, as measured by the coefficient of variation, decreases slightly. Thus the standard deviation is only appropriate as a measure of variability when the mean level of the determinations is fairly constant. The coefficient of variation is however appropriate when the mean level of the determinations is fairly variable providing that the highest mean level is not more than approximately three times the lowest level.

Solution to Problem 2

(a)

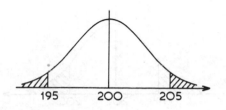

Percentage of determinations above 205:

$$\text{Standardized value} = (205 - 200)/2.0 = 2.500$$

$$\text{Percentage} = 0.62\%$$

211

Percentage of determinations below 195:

$$\text{Standardized value} = (195 - 200)/2.0 = -2.500$$
$$\text{Percentage} \simeq 0.62\%$$

Percentage of batches with which the auto analyser will be declared to be malfunctioning

$$= 0.62 + 0.62$$
$$= 1.24\%$$

(b)

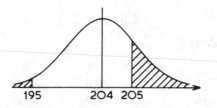

Percentage of determinations above 205:

$$\text{Standardized value} = (205 - 204)/2.0 = 0.500$$
$$\text{Percentage} \simeq 31.0\%$$

Percentage of determinations below 195:

$$\text{Standardized value} = (204 - 195)/2.0 = -4.500$$
$$\text{Percentage} \simeq 0.000\%$$

Percentage of batches with which the auto analyser will be declared to be defective

$$= 31.0\% + 0.000\%$$
$$= 31.0\%$$

(c)

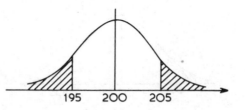

Percentage of determinations above 205:

$$\text{Standardized value} = (205 - 200)/4.0 = 1.250$$
$$\text{Percentage} \simeq 10.5\%$$

Percentage of determinations below 195:

$$\text{Standardized value} = (195 - 200)/4.0 = -1.250$$
$$\text{Percentage} \simeq 10.5\%$$

Percentage of batches with which the auto analyser will be declared to be malfunctioning

$$= 10.5 + 10.5$$
$$= 21.0\%$$

(d)

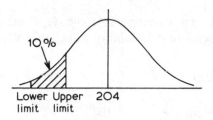

Lower Upper 204
limit limit

We shall initially only consider the upper limit, the area below the lower limit being negligible.

The standardized value giving a left-hand tail of 10% is -1.282.

$$\text{Standardized value} = (\text{value} - \text{mean})/\text{SD}$$
$$-1.282 = (\text{value} - 204)/2$$
$$\text{Value} = 201.44$$

Thus the upper limit is 201.44 and the lower limit is $(200 - 1.44) = 198.56$.

(e) With the present limits the control procedure is far from certain of detecting a bias or lack of precision. If the limits are altered as given in part (d) it will result in a high probability that the auto analyser is wrongly declared to be malfunctioning when it is working correctly.

The one way of improving the procedure is to use more than one control sample. Using the mean of several control samples will enable a better check to be made for possible bias while the standard deviation of the control samples can be used to check the precision.

Chapter 3

Solution to Problem 1

$$\text{SD} = 1.225$$
$$r = 1.96\sqrt{(2)}1.225$$
$$= 3.4$$

(a) The customer would seem to have a valid criticism; it is essential that the operator is unaware that certain samples are replicates. This will be difficult to accomplish with ten replicates in a single batch. A better method would be to use three or four replicates per batch and calculate a combined standard deviation as outlined in Chapter 6.

(b) The customer is incorrect. Repeatability refers to two determinations made under 'repeatability conditions' which implies that the two determinations should be made in the *same* batch, using the same equipment, same operator, etc.

(c) Repeatability refers to the method and not the operator but the customer would seem to have a valid point that several operators should have been used to estimate it.

(d) It is highly likely that a considerable change in mean level will alter the variability and therefore repeatability will change with mean level.

Solution to Problem 2

(a) Coefficient of variation $= \dfrac{\text{SD}}{\text{mean}} \times 100$

	Coefficient of variation	
Powder	VS	D
A	20%	10%
B	10%	10%
C	6.7%	10%

(b)

$$CV_{MS} = \sqrt{(CV_{VS}^2 + CV_D^2)}$$

For powder A
$$= \sqrt{(20^2 + 10^2)}$$
$$= 22.4\%$$

For powder B
$$= \sqrt{(10^2 + 10^2)}$$
$$= 14.1\%$$

For powder C
$$= \sqrt{(6.7^2 + 10^2)}$$
$$= 12.0\%$$

(c) The mean value of MS for each powder is:

Powder	Mean	
A	200/20	= 10.00
B	400/60	= 6.66
C	600/100	= 6.00

$$\text{Standard deviation} = \frac{CV}{100} \times \text{mean}$$

Powder	Standard deviation
A	$10 \times 22.4/100 = 2.24$
B	$6.66 \times 14.1/100 = 0.94$
C	$6.00 \times 12.0/100 = 0.72$

(d) The new standard deviation of *MS* with four tests on *VS* per sample is computed as follows:

Powder	Standard error of VS $= SD/\sqrt{4}$	CV of VS	CV of MS $= \sqrt{(CV_{VS} + CV_D)}$	Mean	SD of MS $= CV \times$ mean/100
A	20	10	14.1	10.00	1.41
B	20	5	11.2	6.66	0.74
C	20	3.3	10.5	6.00	0.63

It can be seen that four tests per sample on *MS* has given a worthwhile reduction in variability to Powder A (2.23→1.41) but not to Powder C (0.72→0.63). This is because the density (*D*) has a higher coefficient of variation (10%) than *VS* (6.7%) with Powder C and it is therefore the test for density that needs several replicates.

Chapter 4

Solution to Problem 1

(a) $n = 8$ $\bar{x} = 50.4$ $s = 0.912$.

Null hypothesis – Green is not biased. (Green's population mean is equal to 50.0.)

Alternative hypothesis – Green is biased. (Green's population mean is not equal to 50.0).

$$\text{Test statistic} - = \frac{|\bar{x} - \mu|}{s/\sqrt{n}} = \frac{|50.4 - 50.0|}{0.912/\sqrt{8}} = 1.24$$

Critical value – from the *t*-table with 7 degrees of freedom for a two-sided test:

2.36 at the 5% significance level
3.50 at the 1% significance level

Decision – We cannot reject the null hypothesis.

Conclusion – There is no evidence that Green is biased. This could however be due to Green's bias being too small to be significant with only eight determinations.

(b)   A confidence interval for Green's population mean is given by:

$$\bar{x} \pm ts/\sqrt{n}$$

where t for a 95% confidence interval $= 2.36$

95% confidence interval is given by:

$$50.4 \pm 2.36 \times 0.912/\sqrt{8}$$
$$50.4 \pm 0.76$$
$$49.64 \quad \text{to} \quad 51.16$$

Therefore the maximum bias is $+1.16$ mg/litre.

(c) $$n \simeq (ts/c)^2$$
$$= (2.36 \times 0.912/0.3)^2$$
$$= 51.4$$

Therefore approximately 52 determinations are needed to estimate Green's bias to within ± 0.3 mg/litre.

Solution to Problem 2

(a)   (i) The width of a 95% confidence interval for the true concentration, using the mean of three determinations and a SD of 0.912 with 7 degrees of freedom, is given by:

$$\pm ts/\sqrt{n}$$
$$\pm 2.36 \times 0.912/\sqrt{3}$$
$$\pm 1.24 \text{ mg/litre}$$

(ii) The variability of corrected determinations is given by:

$$SD = \sqrt{[SD^2(\text{sample}) + SD^2(\text{standard})]}$$
$$= \sqrt{[(0.912)^2 + (0.912)^2]}$$
$$= 1.290$$

The width of a 95% confidence interval for the true concentration using three corrected determinations is:

$$\pm 2.36 \times 1.290/\sqrt{3}$$
$$= 1.76 \text{ mg/litre}$$

To gain benefit from the correction formula, Green's bias would have to be at least $(1.76 - 1.24) = 0.52$ otherwise the accuracy would be decreased.

(b) The variability of mean determinations is given by:

$$SD = \sqrt{[(1/3)^2 (SD^2 + SD^2 + SD^2)]}$$
$$= \frac{SD}{\sqrt{3}}$$
$$= \frac{0.912}{\sqrt{3}}$$
$$= 0.5265$$

The variability of the corrected determinations is given by:

$$SD = \sqrt{[(SD \text{ of standard means})^2}$$
$$+ (SD \text{ of sample means})^2)]$$
$$= \sqrt{[(0.5265)^2 + (0.5265)^2]}$$
$$= 0.7446$$

The width of a 95% confidence interval for the true concentration using the corrected determination is:

$$\pm 2.36(0.7446)/\sqrt{1}$$
$$= \pm 1.76 \text{ mg/litre}$$

This is the same result as that obtained in (a). It does not matter which of the two procedures we use.

Chapter 5

Solution to Problem 1

(a)

True concentration (x)	Measured concentration (y)	$(x-\bar{x})$	$(y-\bar{y})$	$(x-\bar{x})(y-\bar{y})$	$(x-\bar{x})^2$
10	11.0	−25	−25.7	642.5	625
20	21.3	−15	−15.4	231.0	225
30	31.5	−5	−5.2	26.0	25
40	41.9	5	5.2	26.0	25
50	51.9	15	15.2	228.0	225
60	62.6	25	25.9	647.5	625
				1801.0	1750

Slope $(b) = Sxy/Sxx$
$$= 1801/1750 = 1.029$$
$$a = \bar{y} - b\bar{x}$$
$$= 36.7 - 1.029 \ (35.0)$$
$$= 0.68$$

$$y = 0.68 + 1.029x$$

(b)

True concen-tration	Estimated error due to fixed bias (a)	Estimated error due to relative bias (0.29x)	Estimated total error due to bias	Predicted deter-mination	Measured concen-tration (y)	Residual (estimated random error)	Squared residual
10	0.68	0.29	0.97	10.97	11.0	0.03	0.0009
20	0.68	0.58	1.26	21.26	21.3	0.04	0.0016
30	0.68	0.87	1.55	31.55	31.5	-0.05	0.0025
40	0.68	1.16	1.84	41.84	41.9	0.06	0.0036
50	0.68	1.45	2.13	52.13	51.9	-0.23	0.0529
60	0.68	1.74	2.42	62.42	62.6	0.18	0.0324
							0.0939

(c) Residual standard deviation $= \sqrt{(0.0939/4)}$
$$= 0.153$$

(d) 95% confidence interval for the true intercept is given by:

$$a \pm t(ESD)\sqrt{\left(\frac{1}{n} + \frac{\bar{x}^2}{Sxx}\right)}$$

$$0.68 \pm 2.78(0.153)\sqrt{\left[\frac{1}{6} + \frac{(35.0)^2}{1750}\right]}$$

$$0.68 \pm 0.37$$

$$0.31 \quad to \quad 1.05$$

95% confidence interval for the true slope is given by:

$$b \pm t(ESD)/\sqrt{Sxx}$$

$$1.029 \pm 2.78(0.153)/\sqrt{1750}$$

$$1.029 \pm 0.010$$

$$1.019 \quad to \quad 1.039$$

(e) We are 95% certain that the fixed bias is between 0.31 and 1.05 and the relative bias is between 1.9% and 3.9%. We can therefore conclude that both biases are present.

Solution to Problem 2

The narrower confidence interval for the slope is a red herring. Regression through the origin will usually tend to give a smaller confidence interval because of the method of calculation. What matters is the accuracy of the line to the data which is governed by the ESD. Regression through the origin has increased the ESD from 0.153 to 0.354. This is due to the fixed bias being ignored when it is clearly present. In other words, regression through the origin is using a wrong model in relation to the nature of the bias.

Solution to Problem 3

Method (c) gives a good correction at any concentration because the use of the blank corrects for the fixed bias and the use of the standard corrects for the relative bias.

Both methods (a) and (b) give good results if the unknown sample has a concentration equal to that of the standard. However, both methods are found wanting at higher and lower concentrations.

Method (a) compensates for the fixed bias but, at lower concentrations it over-corrects for the relative bias and at higher concentrations it under-corrects. Method (b) on the other hand will always give an excessively low corrected determination when the concentration of the sample is below that of the standard and will give a high result for a sample with concentration above the standard.

Chapter 6

Solution to Problem 1

(a) $s = 0.912$ *degrees of freedom* $= 7$

Using Table F gives $L_1 = 0.66$ and $L_2 = 2.04$ for a 95% confidence interval.

Upper limit $= L_2 \times S = 2.04 \times 0.912 = 1.86$
Lower limit $= L_1 \times S = 0.66 \times 0.912 = 0.60$

Thus we are 95% certain that Green's true standard deviation is between 0.60 and 1.86 mg/litre.

(b) We can check this assumption using Dixon's test. First we put the determinations into ascending order:

x_1	x_2	x_3	x_4	x_5	x_6	x_7	x_8
49.3	49.4	49.8	50.0	50.8	50.8	51.3	51.8

Null hypothesis – All determinations came from the same normal distribution.

Alternative hypothesis – Either the highest determination or the lowest determination did not come from the same normal distribution as the other seven.

Test statistic – = the greater of A or B.

$$A = \frac{x_2 - x_1}{x_7 - x_1} \qquad B = \frac{x_8 - x_7}{x_8 - x_2}$$

$$= \frac{49.4 - 49.3}{51.3 - 49.3} \qquad = \frac{51.8 - 51.3}{51.8 - 49.4}$$

$$= 0.050 \qquad = \underline{0.208}$$

Critical values – from Table G for a sample size of 8 are:

   0.608 at the 5% significance level
   0.717 at the 1% significance level

Decision – We cannot reject the null hypothesis.

Conclusions – We are unable to conclude that the assumption is violated.

Solution to Problem 2

(a) The combined variance $= [3(0.411)^2 + 3(0.793)^2 \ldots + 2(0.585)^2 + 3(0.780)^2]/$
$$(3+3 \ldots 2+3)$$
$$= 8.354/20$$
$$= 0.4177$$

   Combined SD $= \sqrt{0.4177}$
$$= 0.646$$

(b) *Null hypothesis* – The precision of Green is equal to that for the experienced operators $(\sigma_G = \sigma_E)$.

   Alternative hypothesis – Green is less precise than the experienced operators $(\sigma_G > \sigma_E)$.

   Test statistic $- = \dfrac{\text{variance for Green}}{\text{variance for experienced operators}}$

$$= \frac{(0.912)^2}{(0.646)^2} = 1.99$$

   Critical values – from the one-sided F-table with 7 and 20 degrees of freedom:

   2.49 at the 5% significance level
   3.70 at the 1% significance level

   Decision – We cannot reject the null hypothesis.

   Conclusion – There is no evidence that Green is less precise than the experienced operators.

(c) The F-value with 7 and 20 degrees of freedom is 2.49 at a 5% significance level. Therefore

$$1 + \frac{x}{100} = 2.49$$

$$x = 149\%$$

Therefore Green will have to be 150% more variable before the F-test would be likely to detect it.

Chapter 7

Solution to Problem 1

(a)
$$b = Sxy/Sxx$$
$$= 332\,400/825\,000$$
$$= 0.4029$$
$$a = \bar{y} - b\bar{x}$$
$$= 242 - 0.4029 \times 550$$
$$= 20.4$$
$$y = 20.4 + 0.4029x$$

(b) Residual sum of squares $= Syy - b^2 Sxx$
$$= 134\,260 - (0.4029)^2(825\,000)$$
$$= 339.1$$

(c) Residual standard deviation $= \sqrt{\left(\dfrac{\text{residual sum of squares}}{\text{degrees of freedom}}\right)}$
$$= \sqrt{(339.1/8)}$$
$$= 6.51 \text{ microamps}$$

(d) The 95% confidence interval for a single determination is given by:

$$\left(\frac{Y-a}{b}\right) \pm \frac{t(\text{ESD})}{b} \sqrt{\left[1 + \frac{1}{n} + \left(\frac{Y-\bar{y}}{b}\right)^2 \Big/ Sxx\right]}$$

$$= \left(\frac{400 - 20.4}{0.4029}\right) \pm \frac{2.31 \times 6.51}{0.4029} \sqrt{\left[1 + \frac{1}{10} + \left(\frac{400 - 242}{0.4029}\right)^2 \Big/ 825\,000\right]}$$

$$= 942.1 \pm 42.3 \text{ p.p.m.}$$

i.e. 900 to 984 p.p.m.

(e) Judging from the confidence interval in the previous problem, the width of a 95% confidence interval at a concentration of 0.1% (1000 p.p.m.) will be approximately 45 p.p.m. Therefore a sample with true concentration of 1000 might appear to have a concentration as low as 955 p.p.m.

$$Y = 20.4 + 0.4029x$$
$$Y = 20.4 + 0.4029 \times 955$$
$$Y = 405 \text{ microamps}$$

Thus the microamp reading should be below 405 to ensure that the concentration is less than 0.1%.

Solution to Problem 2

(a) Slope (b)
$$= \frac{\sum xz}{\sum x^2}$$

$$= \frac{1\,553\,400}{3\,850\,000} = 0.403\,480\,5$$

$$y = 0.403\,480\,5z$$

(b) Residual sum of squares $= \sum z^2 - b^2 \sum x^2$
$$= 627\,100 - (0.403\,480\,5)^2\,(3\,850\,000)$$
$$= 333.36$$

$$\mathrm{RSD} = \sqrt{\left(\frac{333.36}{9}\right)}$$

$$= 6.086$$

Error standard deviation $= \mathrm{RSD}/1.1 = 5.532$

(c) $Z = 420 - 40 = 380$

95% confidence interval for true concentration is given by:

$$\frac{Z}{b} \pm \frac{t(\mathrm{ESD})}{b} \sqrt{\left[2 + \frac{Z^2}{b^2 \sum x^2}\left(1 + \frac{(\sum x)^2}{\sum x^2}\right)\right]}$$

$$= \frac{380}{0.403\,480\,5} \pm \frac{2.26 \times 5.532}{0.403\,480\,5} \sqrt{\left[2 + \frac{(380)^2}{(0.403\,480\,5)^2 \times 3\,850\,000}\left(1 + \frac{(5500)^2}{3\,850\,000}\right)\right]}$$

$$= 942 \pm 62$$

(d) The subtraction of a blank has removed the batch to batch fixed bias. Clearly if a fixed bias is present in each batch in differing quantities it is essential to remove it even at the cost of increasing the confidence interval.

Solution to Problem 3

(a)   Slope $(b) = \dfrac{\text{standard reading} - \text{blank reading}}{1000}$

$$= \frac{429 - 18}{1000}$$

$$= 0.411$$

(b) $\sum z = 1000 \quad \sum z^2 = (1000)^2 = 1\,000\,000$

(c) $$Z = 217 - 18 = 199$$

95% confidence interval is given by:

$$\frac{Z}{b} + \frac{t(\text{ESD})}{b}\sqrt{\left[2 + \frac{Z^2}{b^2\sum z^2}\left(1 + \frac{(\sum z)^2}{\sum z^2}\right)\right]}$$

$$= \frac{199}{0.411} + \frac{1.96 \times 5.5}{0.411}\sqrt{\left[2 + \frac{(199)^2}{(0.411)^2 \times 1\,000\,000}\left(1 + \frac{(1000)^2}{1\,000\,000}\right)\right]}$$

$$= 484 \pm 41$$

i.e. 443 to 525 p.p.m.

The confidence interval includes the true concentration of 500 p.p.m.

Solution to Problem 4

A confidence interval for the true concentration of the unknown sample is given by:

$$\frac{\bar{Y} - a}{b} \pm \frac{t(\text{ESD})}{b}\sqrt{\left[\frac{1}{m} + \frac{1}{n} + \left(\frac{\bar{Y} - \bar{y}}{b}\right)^2 \Big/ \text{Sxx}\right]}$$

where $\bar{y}$ = mean of n absorbance readings in the calibration experiment;
$\bar{Y}$ = mean of m absorbance readings on the unknown sample.

If the true concentration is close to 50 p.p.m. then $\bar{Y}$ will be approximately equal to $\bar{y}$ and we can neglect the third term under the square root. The half width of the interval is then given by:

$$\text{HW} = \frac{t(s)}{b}\sqrt{\left(\frac{1}{m} + \frac{1}{n}\right)}$$

where s is an estimate of the error SD.

(a) Estimating the error SD from five standards at 50 p.p.m.

$$n = 4 \quad m = 1 \quad t_4 = 2.78$$
$$\text{HW} = 3.108 \ (s/b)$$

(b) Using the RSD from the calibration to estimate error SD

$$n = 9 \quad m = 1 \quad t_7 = 2.36$$
$$\text{HW} = 2.488 \ (s/b)$$

(c) Using the RSD from the calibration to estimate error SD

$$n = 6 \quad m = 4 \quad t_4 = 2.78$$
$$\text{HW} = 1.794 \ (s/b)$$

(d) Using the RSD from the calibration to estimate the error SD

$$n=3 \quad m=7 \quad t_1 = 12.71$$
$$HW = 8.771 \; (s/b)$$

Of the four strategies considered, (c) can be expected to give the narrowest confidence interval. Within the constraint of only sufficient reagent for ten determinations we can improve on strategy (c). By using seven standards (0, 17, 34, 50, 67, 84 and 100 p.p.m.) in a calibration experiment and making three determinations on the unknown sample we get a half width for the confidence interval of 1.773 (s/b).

Chapter 8

Solution to Problem 1

(a)

Concentration (p.p.m.)	10	40	70	100
Standard deviation (s)	1.203	1.821	2.439	3.057

(b)

Concentration (p.p.m.)	10	40	70	100	Total
$(1/s^2)$	0.6910	0.3016	0.1681	0.1070	1.2677
Weight (w)	2.180	0.952	0.530	0.338	4.000

(c)

	x	y	w	wx	wy	wxy	wx^2	wy^2
	10	5	2.180	21.80	10.900	109.00	218.0	54.500
	40	17	0.952	38.08	16.184	647.36	1523.2	275.128
	70	35	0.530	37.10	18.550	1298.50	2597.0	649.250
	100	45	0.338	33.80	15.210	1521.00	3380.0	684.450
Total	220	102	4.000	130.78	60.844	3575.86	7718.2	1663.328
Mean	55.0	25.5	1.000	32.695	15.211			

$$Swxy = \sum wxy - n\bar{x}_w \bar{y}_w$$
$$= 3575.86 - 4(32.695)15.211$$
$$= 1586.565$$
$$Swxx = \sum wx^2 - n(\bar{x}_w)^2$$
$$= 7718.2 - 4(32.695)^2$$
$$= 3442.348$$

$b = Swxy/Swxx$

$\qquad = 1586.565/3442.348$

$\qquad = 0.460\,896$

$a = \bar{y}_w - b\bar{x}_w$

$\qquad = 15.211 - 0.460\,896(32.695)$

$\qquad = 0.1420$

The equation of the weighted least squares regression line is $y = 0.142 + 0.461x$.

Using ordinary least squares we would get the equation $y = 0.200 + 0.460x$.

(d) $Swyy = \sum wx^2 - n(\bar{y}_w)^2$

$\qquad = 1663.328 - 4(15.211)^2$

$\qquad = 737.830$

Residual standard deviation

$\qquad = \sqrt{[(Swyy - b^2 Swxx)/(n-2)]}$

$\qquad = \sqrt{[(737.830 - (0.460\,896)^2 3442.348)/2]}$

$\qquad = 1.8150$

(e) A confidence interval for the true concentration of an unknown sample is given by:

$$\left(\frac{Y-a}{b}\right) \pm \frac{t(\text{ESD})}{b} \sqrt{\left[\frac{1}{w} + \frac{1}{n} + \left(\frac{Y-y_w}{b}\right)^2 \Big/ Swxx\right]}$$

where Y is the measured absorbance of the unknown sample. We will substitute $a = 0.142$, $b = 0.461$, $t_2 = 4.30$, $\text{ESD} = 1.8150$, $n = 4$, $\bar{y}_w = 15.211$, $Swxx = 3442.348$

(i) When $Y = 5$

$$x = (5 - 0.142)/0.461 = 10.538$$

$$\left(\begin{array}{c}\text{SD of repeat}\\ \text{determinations}\end{array}\right) = 0.997 + 0.206x$$

$$= 0.997 + 0.0206(10.538)$$

$$= 1.2141$$

$$w = n(1/s^2)/(\text{total of } 1/s^2 \text{ in calibration experiment})$$

$$= 4(1/(1.2141)^2)/(1.2677)$$

$$= 2.141$$

The 95% confidence interval for the true concentration of a sample that gave an absorbance reading of 5 is:

$$\left(\frac{5-0.142}{0.461}\right) \pm \frac{4.30(1.8150)}{0.461} \sqrt{\left[\frac{1}{2.141} + \frac{1}{4} + \left(\frac{5-15.211}{0.461}\right)^2 \middle/ 3442.348\right]}$$

$$= 10.538 \pm 16.930\sqrt{(0.467+0.250+0.143)}$$

$$= 10.538 \pm 15.700 \text{ p.p.m.}$$

The confidence interval that would have been obtained using ordinary regression is 10.435 ± 28.25.

(ii) When $Y=45$

$$x = (45-0.142)/0.461 = 97.306$$

$$\left(\begin{array}{c}\text{SD of repeat}\\\text{determinations}\end{array}\right) = 0.997 + 0.0206x$$

$$= 0.997 + 0.0206(97.306)$$

$$= 3.0015$$

$$w = n(1/s^2)/(\text{total of } 1/s^2 \text{ for calibration experiment})$$

$$= 4(1/(3.0015)^2)/(1.2677)$$

$$= 0.3502$$

The 95% confidence interval for the true concentration of a sample that gave an absorbance reading of 45 is:

$$\left(\frac{45-0.142}{0.461}\right) \pm \frac{4.30(1.8150)}{0.461} \sqrt{\left[\frac{1}{0.3502} + \frac{1}{4} + \left(\frac{45-15.211}{0.461}\right)^2 \middle/ 3442.348\right]}$$

$$= 97.306 \pm 16.930\sqrt{(2.855+0.250+1.213)}$$

$$= 97.306 \pm 35.181 \text{ p.p.m.}$$

The confidence interval that would have been obtained using ordinary regression is 97.391 ± 27.897.

Note: The above confidence intervals are very wide because we are estimating the residual standard deviation from the results of the calibration experiment. Because only four points were used we have only 2 degrees of freedom for this residual. Having already carried out an extensive experiment in order to investigate the error structure, we could obtain a much better estimate of residual variation from this earlier experiment. We simply reverse our earlier calculation to find a standard deviation corresponding to a weight of 1.000.

$$w = n(1/s^2)/(\text{total of } 1/s^2 \text{ in calibration experiment})$$

$$1 = 4(1/s^2)/(1.2677)$$

$$s = 1.776$$

Using ESD=1.776 with infinite degrees of freedom we would get a much narrower confidence interval for the true concentration of an unknown sample.

Solution to Problem 2

Step 1 : Draw a scatter diagram. Though the question does not ask you to draw a diagram you will probably feel very uneasy making decisions about individual points without consulting one. The diagram below certainly gives the impression that the linear region ends at 16 p.p.m., or 18 p.p.m. perhaps.

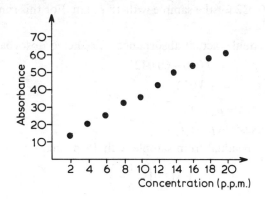

Step 2: Using the equation fitted over the range 2 to 18 p.p.m. inclusive. Substituting $x=20$ into the equation, $y=9.389+2.750x$, gives a predicted absorbance of 64.389 for the sample with 20 p.p.m. For this sample we can now calculate a residual.

$$\text{Residual} = \text{actual absorbance} - \text{predicted absorbance}$$
$$= 61.000 - 64.389$$
$$= -3.389$$

We would expect a negative residual for this point if it were beyond the limit of linearity. Could such a large negative residual have arisen by chance if the linear region extended beyond 20 p.p.m.

Null hypothesis – $\mu_A = \mu_B = 0$

Alternative hypothesis – $\mu_A < \mu_B$

Test statistic – $= \dfrac{|\text{residual from sample with 20 p.p.m.}|}{\text{RSD}\sqrt{\left(\dfrac{n+1}{n}\right)}}$

$$= \frac{3.389}{1.409\sqrt{(10/9)}}$$
$$= 2.28$$

Crticial values – from the *t*-table with 7 degrees of freedom for a one-sided test:

> 1.89 at the 5% significance level
> 3.00 at the 1% significance level

Decision – We reject the null hypothesis at the 5% level of significance.

Conclusion – We conclude that a concentration of 20 p.p.m. is beyond the limit of linearity.

Step 3: Using the equation fitted over the range 2 to 16 p.p.m. inclusive. Substituting $x = 18$ into the equation, $y = 8.714 + 2.851x$, gives a predicted absorbance of 60.032 for the sample with 18 p.p.m. For this sample we can now calculate a residual.

> Residual = actual absorbance – predicted absorbance

$$= 57.000 - 60.032$$

$$= -3.032$$

Null hypothesis – $\mu_A = \mu_B = 0$

Alternative hypothesis – $\mu_A < \mu_B$

$$\text{Test statistic} - = \frac{|\text{residual from sample with 18 p.p.m.}|}{\text{RSD}\sqrt{\left(\dfrac{n+1}{n}\right)}}$$

$$= \frac{3.032}{1.166\sqrt{(9/8)}}$$

$$= 2.45$$

Critical values – from the *t*-table with 6 degrees of freedom for a one-sided test:

> 1.94 at the 5% significance level
> 3.14 at the 1% significance level

Decision – Reject the null hypothesis at the 5% level of significance.

Conclusion – We conclude that a concentration of 18 p.p.m. is beyond the limit of linearity.

Step 4: Using the equation fitted over the range 2 to 14 p.p.m. inclusive. Substituting $x = 16$ into the equation, $y = 8.571 + 2.875x$, gives a predicted absorbance of 54.571 for the sample with 16 p.p.m. For this sample we can now calculate a residual.

> Residual = actual absorbance – predicted absorbance

$$= 54.000 - 54.571$$

$$= -0.571$$

Null hypothesis – $\mu_A = \mu_B = 0$

Alternative hypothesis – $\mu_A < \mu_B$

$$\text{Test statistic} - = \frac{\left|\text{residual from the sample with 16 p.p.m.}\right|}{\text{RSD}\sqrt{\left(\dfrac{n+1}{n}\right)}}$$

$$= \frac{0.571}{1.262\sqrt{(8/7)}}$$

$$= 0.42$$

Critical values – from the *t*-table with 5 degrees of freedom for a one-sided test:

2.02 at the 5% significance level

3.36 at the 1% significance level

Decision – We cannot reject the null hypothesis.

Conclusion – We are unable to conclude that 16 p.p.m. is beyond the range of linearity.

After carrying out the above significance tests and examining the scatter diagram, it is reasonable to conclude that the limit of the linear range is beyond 16 p.p.m.

Chapter 9

Solution to Problem 1

(a) Analysis of variance table

Source	Sum of squares	Degrees of freedom	Mean square
Between-laboratories	1.00	3	0.3333
Within-laboratories	0.68	16	0.0425
Total	1.68	19	

Parts (i) to (iv) see table above.
(v) See table overleaf.
(vi) See table above.

(b) The within laboratories standard deviation (σ_w) is the square root of within laboratories mean square and is equal to 0.206 $(\sqrt{0.0425})$.

(c) The between laboratories standard deviation (σ_b^2) is given by the square root of:

$$\frac{\text{between laboratories mean square} - \text{within laboratories mean square}}{\text{no. of determinations per laboratory}}$$

$$\sqrt{\left(\frac{0.3333 - 0.0425}{5}\right)} = \sqrt{0.0582} = 0.241$$

(v)

1	2	3	4	5	6	7	8	9	10
Laboratory	Determination	Determination minus overall mean	Column 3 squared	Laboratory mean	Determination minus lab mean	Column 6 squared	Determination replaced by lab mean	Lab mean minus overall mean	Column 9 squared
A	0.8	−0.3	0.09		−0.2	0.04	1.0	−0.1	0.01
	1.1	0.0	0.00		0.1	0.01	1.0	−0.1	0.01
	1.2	0.1	0.01	1.0	0.2	0.04	1.0	−0.1	0.01
	0.8	−0.3	0.09		−0.2	0.04	1.0	−0.1	0.01
	1.1	0.0	0.00		0.1	0.01	1.0	−0.1	0.01
B	0.7	−0.4	0.16		−0.1	0.01	0.8	−0.3	0.09
	1.1	0.0	0.00		0.3	0.09	0.8	−0.3	0.09
	0.5	−0.6	0.36	0.8	−0.3	0.09	0.8	−0.3	0.09
	0.9	−0.2	0.04		0.1	0.01	0.8	−0.3	0.09
	0.8	−0.3	0.09		0.0	0.00	0.8	−0.3	0.09
C	1.6	0.5	0.25		0.2	0.04	1.4	0.3	0.09
	1.3	0.2	0.04		−0.1	0.01	1.4	0.3	0.09
	1.1	0.0	0.00	1.4	−0.3	0.09	1.4	0.3	0.09
	1.6	0.5	0.25		0.2	0.04	1.4	0.3	0.09
	1.4	0.3	0.09		0.0	0.00	1.4	0.3	0.09
D	1.4	0.3	0.09		0.2	0.04	1.2	0.1	0.01
	1.0	−0.1	0.01		−0.2	0.04	1.2	0.1	0.01
	1.4	0.3	0.09	1.2	0.2	0.04	1.2	0.1	0.01
	1.2	0.1	0.01		0.0	0.00	1.2	0.1	0.01
	1.0	−0.1	0.01		−0.2	0.04	1.2	0.1	0.01
Total	22.0	0.0	1.68			0.68			1.00
Mean	1.1		Total SS			Within labs SS			Between labs SS

(d) (i) the repeatability of the test method $(r) = t_{16}\sqrt{(2)}\sqrt{(\text{estimate of } \sigma_w^2)}$

$$= 2.12\sqrt{(2)}\sqrt{(0.0425)}$$

$$0.618$$

(e) The reproducibility of the test method (R)

$$= t_7\sqrt{(2)}\sqrt{[(\text{estimate of } \sigma_w^2)}$$
$$+ (\text{estimate of } \sigma_b^2)]$$

$$= 2.36\sqrt{(2)}\sqrt{(0.0425 + 0.0582)}$$

$$= 1.059$$

(f) *Null hypothesis* – $\sigma_b^2 = 0$ no difference in mean ammonia assay between laboratories.

Alternative hypothesis – $\sigma_b^2 > 0$ there is a difference between laboratories.

Test statistic – 0.3333/0.0425

$$= 7.84$$

Critical values – 3.24 at 5% and 5.29 at 1% using F-tables, one-sided risk, with 3 and 16 degrees of freedom.

We therefore reject the null hypothesis at a 1% significance level and conclude the laboratories give significantly different mean results.

(g) The Cochran test statistic is:

$$\frac{(0.224)^2}{(0.187)^2 + (0.224)^2 + (0.212)^2 + (0.200)^2}$$

$$= 0.295$$

Referring to Table 13 gives critical values with $n = 5$ and $p = 4$ of 0.721 (1%) and 0.629 (5%). Clearly we can assume the variances are homogeneous.

(h) Dixon's test:

Null hypothesis – All four means come from the same normal distribution.

Alternative hypothesis – Either the highest or lowest mean does not come from the same normal distribution as the other three.

Test statistic – The greater of A or B:

x_1	x_2	x_3	x_4
24.8	25.0	25.2	25.4

$$A = \frac{x_2 - x_1}{x_4 - x_1} \qquad B = \frac{x_4 - x_3}{x_4 - x_1}$$

$$A = \frac{0.2}{0.6} \qquad B = \frac{0.2}{0.6}$$

$$= 0.33 \qquad = 0.33$$

Test statistic $= 0.33$

Critical values – 0.829 at the 5% significance level
                  0.926 at the 1% significance level

   Decision – We cannot reject the null hypothesis.

   Conclusion – There is no evidence that a laboratory is an outlier regarding
      mean determination level.

(i) A 95% confidence interval for the true mean determination:

$$\text{overall mean} \pm t_7 \sqrt{\left(\frac{\text{estimate of } \sigma_b^2}{\text{no. of labs}} + \frac{\text{estimate of } \sigma_w^2}{\text{total no. of determinations}} \right)}$$

$$= 25.1 \pm 2.36 \sqrt{\left(\frac{0.0582}{4} + \frac{0.0425}{20} \right)}$$

$$= 25.1 \pm 0.30$$

$$= 24.80 \text{ to } 25.40$$

Since the true concentration is 25.0, the maximum bias is +0.40.

(j) A confidence interval for true mean determination of laboratory C is given
   by:

$$\text{lab mean} \pm t \sqrt{\left(\frac{\text{estimate of } \sigma_w^2}{\text{no. of determinations in that laboratory}} \right)}$$

   t with 16 d.f. is 2.12

$$= 25.4 \pm 2.12 \sqrt{\left(\frac{0.0425}{5} \right)}$$

$$= 25.4 \pm 0.20$$

   i.e. 25.20   to   25.60

Therefore the maximum bias is +0.60.

Solution to Problem 2

(a) It is highly likely that laboratories will have different biases and these will give
    rise to similar patterns as those observed in determinations from laboratory
    C. Dixon's test can be used to check whether the mean level of a laboratory
    should be classed as an outlier. In this case the test does not reveal any
    outliers.

(b) If we require an estimate of the repeatability, determinations should be
    carried out in the same batch to comply with conditions given in BS 5497.

(c) This is a possibility that can be avoided by having five samples with
    marginally different concentrations. This result is a slightly different analysis
    and will be covered in Chapter 11.

(d) Cochran's test shows that the variances are not significantly different and it is
    therefore reasonable to combine them to give one estimate of the standard
    deviation.

Chapter 10

Solution to Problem 1

(a) SD of cell means $= 0.698\,049\,229$

Between cells sum of squares $= 11(0.698\,049\,229)^2 3$

$$= 16.08$$

Source of variation	Sum of squares	Degrees of freedom	Mean square
Between-cells	16.08	11	1.462
Within-cells	2.80	24	0.117
Total	18.88	35	

(b)

SD of day means $= 0.476\,095\,228$

Between-days sums of squares $= 3(0.476\,095\,228)^2 9$

$$= 6.12$$

SD of operator means $= 0.608\,276\,253$

Between-operators sum of squares $= 2(0.608\,276\,253)^2 12$

$$= 8.88$$

Source of variation	Sum of squares	Degrees of freedom	Mean square
Between-days	6.12	3	2.040
Between-operators	8.88	2	4.440
Interaction	1.08	6	0.180
Residual	2.80	24	0.117
Total	18.88	35	

(c) *Interaction v. residual*

Test statistic $= 0.180/0.117 = 1.54$

Critical values (6, 24 degrees of freedom) 5% level $= 2.51$

1% level $= 3.67$

We can conclude that there is no evidence of a variable bias and therefore the main effects are tested against the residual and not the interaction.

Between-operators v. residual

$$\text{Test statistic} = 4.44/0.117 = 37.94$$

Critical values (2, 24 degrees of freedom) 5% level = 3.40

1% level = 5.61

Clearly there is evidence of a variation in bias from operator to operator.

Between days v. residual

$$\text{Test statistic} = 2.04/0.117 = 17.43$$

Critical values (3, 24 degrees of freedom) 5% level = 3.01

1% level = 4.72

Clearly there is evidence of a significant day to day variation in bias.

(d)   Testing variance = 0.117

$$\text{Operators variance} = \frac{4.440 - 0.117}{12} = 0.360\,25$$

$$\text{Day-to-day variance} = \frac{2.040 - 0.117}{9} = 0.213\,67$$

$$\text{SD of single determinations} = \sqrt{\left(\begin{array}{c}\text{testing}\\\text{variance}\end{array}\right) + \left(\begin{array}{c}\text{operator}\\\text{variance}\end{array}\right) + \left(\begin{array}{c}\text{day-to-day}\\\text{variance}\end{array}\right)}$$

$$\text{SD of a single determination} = \sqrt{(0.117 + 0.360\,25 + 0.213\,67)}$$
$$= 0.831$$

(e) The duplicate readings should be independent of each other regarding the three sources of variation. In other words they should use different operators on different days.

This will give:

SD of mean determinations

$$= \sqrt{\left(\frac{\text{testing variance}}{2}\right) + \left(\frac{\text{operator variance}}{2}\right) + \left(\frac{\text{day-to-day variance}}{2}\right)}$$

$$= \sqrt{\left(\frac{0.117}{2} + \frac{0.360\,25}{2} + \frac{0.2136}{2}\right)}$$

$$= 0.588$$

Note: If duplicates are required which are as close as possible to each other they should be carried out by the same operator on the same day. However these do not give as good an estimate of the true mean as the above method. It should be noted that the above method is no substitute for removing the inefficient (biased) operator.

Chapter 11

Solution to Problem 1

(a) Since there appears to be only two methods and both are included in the experiment, 'methods' calls for a complete model. On the other hand, 'operators' calls for a sampling model, since the four operators are a sample of those who could use the methods either now or in the future.

(b)

Source of variation	Degrees of freedom
Due to methods	1
Due to operators	3
Interaction (operator × methods)	3
Residual	24
Total	31

(c) *Laboratory A:* Since the interaction is not significant when compared with the residual we test *both* operators and methods against the residual. We conclude that methods do not give significantly different mean levels but operators do give different levels. Tests 4 and 5 should not have been carried out.

Laboratory B: The interaction is significant compared with the residual and we therefore test operators against the interaction which gives operators to be not significant. We can therefore conclude the operators do not have a fixed bias but do have a bias which is different for each method. The methods mean square must be tested against the residual, regardless of the decision reached in test no. 3, and we can conclude that the methods do give significantly different mean levels. Tests 2 and 4 should not have been carried out.

Solution to Problem 2

(a) The variables are:
Types of beer   S or L
Laboratories   seven in all
Different months within type of beer
Different batches within each month

(b) Although we have not enough information to make a final judgement it would appear that type of beer is a complete variable and that the others are sampling variables.

(c) 'Months within type of beer' and 'batch within month' are both nested variables.

(d)

Sources of variation	Degrees of freedom
Between-beers	1
Between-laboratories	6
Interaction (B × L)	6
Between months within beers	2
Between batches within months	4
Residual	36
Total	55

The residual *could* be broken down into two components:

 (i) Interaction (L × (months within beers)) with 12 d.f.

(ii) Interaction (L × (batches within months)) with 24 d.f.

Bibliography

Barnett, V. and Lewis, T. (1978) *Outliers in Statistical Data*, Wiley.

BS 5497 (1979) *Precision of test methods*, Part 1, *Guide to the determination of repeatability and reproducibility of a standard test method*, British Standards Institute.

Brownlee, K. A. (1960) *Statistical Theory and Methodology in Science and Engineering*, 1st edn, Wiley.

Caulcutt, R. (1983) *Statistics in Research and Development*, Chapman and Hall, London.

Davies, O. L. (Ed.) (1978) *Design and Analysis of Industrial Experiments*, Longmans.

Davies, O. L. & Goldsmith, P. L. (1972) *Statistical Methods in Research and Production*, Longmans.

Eckschlager, K. (1969) *Errors, Measurement and Results in Chemical Analysis*, Van Nostrand.

IS 4259 (1979) *Petroleum products – determination and application of precision data in relation to methods of test*, International Organisation for Standardisation.

Johnson, N. and Leone, F. C. (1964) *Statistics and Experimental Design*, Volume 2, Wiley.

Liteanu, C. and Rica, I. (1980) *Statistical Theory and Methodology of Trace Analysis*, Ellis Horwood.

Massart, D. L. *et al.* (1978) *Evaluation and Optimisation of Laboratory Methods and Analytical Procedures*, Elsevier.

Nalimov, V. V. (1963) *The Application of Mathematical Statistics to Chemical Analysis*, Pergamon.

Pearson, E. S. and Hartley, H. O. (1970) *Biometrika Tables for Statisticians*, Cambridge University Press.

Smith, R. and James, G. V. (1981) *The Sampling of Bulk Materials*, The Royal Society of Chemistry.

Youden, W. J. and Steiner, E. H. (1975) *Statistical Manual of the Association of Official Analytical Chemists*, AOAC.

Wilson, A. L. (1979) Approach for achieving comparable analytical results from a number of laboratories. *The Analyst*, **104**, 273–89.

Statistical tables

Table A The normal distribution
[standardized value = (value − mean)/SD]

Percentage exceeding the value — Mean, Value

Standardized value	% exceeding the value	Standardized value	% exceeding the value	Standardized value	% exceeding the value	Standardized value	% exceeding the value	Standardized value	% exceeding the value	Standardized value	% exceeding the value	Standardized value	% exceeding the value
0.000	50.0	0.842	20.0	1.645	5.0	2.054	2.00	2.575	0.50	2.877	0.200	3.287	0.050
0.025	49.0	0.860	19.5	1.655	4.9	2.064	1.95	2.582	0.49	2.885	0.195	3.317	0.045
0.050	48.0	0.878	19.0	1.664	4.8	2.075	1.90	2.589	0.48	2.893	0.190	3.349	0.040
0.075	47.0	0.896	18.5	1.675	4.7	2.086	1.85	2.597	0.47	2.901	0.185	3.385	0.035
0.101	46.0	0.915	18.0	1.685	4.6	2.097	1.80	2.604	0.46	2.910	0.180	3.427	0.030
0.126	45.0	0.935	17.5	1.695	4.5	2.108	1.75	2.612	0.45	2.919	0.175	3.476	0.025
0.151	44.0	0.954	17.0	1.706	4.4	2.120	1.70	2.619	0.44	2.928	0.170	3.534	0.020
0.176	43.0	0.974	16.5	1.717	4.3	2.132	1.65	2.627	0.43	2.937	0.165	3.607	0.015
0.202	42.0	0.994	16.0	1.728	4.2	2.144	1.60	2.635	0.42	2.946	0.160	3.707	0.010
0.228	41.0	1.015	15.5	1.739	4.1	2.157	1.55	2.643	0.41	2.956	0.155	3.869	0.005
0.253	40.0	1.036	15.0	1.751	4.0	2.170	1.50	2.652	0.40	2.966	0.150		
0.279	39.0	1.058	14.5	1.762	3.9	2.183	1.45	2.660	0.39	2.977	0.145		
0.305	38.0	1.080	14.0	1.774	3.8	2.197	1.40	2.669	0.38	2.987	0.140		
0.332	37.0	1.103	13.5	1.786	3.7	2.211	1.35	2.678	0.37	2.998	0.135		
0.358	36.0	1.126	13.0	1.799	3.6	2.226	1.30	2.687	0.36	3.010	0.130		
0.385	35.0	1.150	12.5	1.812	3.5	2.241	1.25	2.696	0.35	3.022	0.125		
0.412	34.0	1.175	12.0	1.825	3.4	2.257	1.20	2.706	0.34	3.034	0.120		
0.440	33.0	1.200	11.5	1.838	3.3	2.273	1.15	2.716	0.33	3.047	0.115		
0.468	32.0	1.226	11.0	1.852	3.2	2.290	1.10	2.726	0.32	3.060	0.110		
0.496	31.0	1.254	10.5	1.866	3.1	2.308	1.05	2.736	0.31	3.074	0.105		
0.524	30.0	1.282	10.0	1.881	3.0	2.326	1.00	2.747	0.30	3.089	0.100		
0.553	29.0	1.311	9.5	1.896	2.9	2.345	0.95	2.758	0.29	3.104	0.095		
0.583	28.0	1.341	9.0	1.911	2.8	2.365	0.90	2.770	0.28	3.120	0.090		
0.613	27.0	1.372	8.5	1.927	2.7	2.386	0.85	2.781	0.27	3.136	0.085		
0.643	26.0	1.405	8.0	1.943	2.6	2.409	0.80	2.794	0.26	3.154	0.080		
0.674	25.0	1.439	7.5	1.960	2.5	2.432	0.75	2.806	0.25	3.172	0.075		
0.706	24.0	1.476	7.0	1.977	2.4	2.457	0.70	2.819	0.24	3.152	0.070		
0.739	23.0	1.514	6.5	1.995	2.3	2.483	0.65	2.833	0.23	3.214	0.065		
0.772	22.0	1.555	6.0	2.014	2.2	2.512	0.60	2.847	0.22	3.237	0.060		
0.806	21.0	1.598	5.5	2.033	2.1	2.542	0.55	2.862	0.21	3.261	0.055		

Degrees of freedom	Two-sided test 10% (0.10)	5% (0.05)	1% (0.01)	One-sided test 10% (0.10)	5% (0.05)	1% (0.01)
1	6.31	12.71	63.66	3.08	6.31	31.82
2	2.92	4.30	9.92	1.89	2.92	6.97
3	2.35	3.18	5.84	1.64	2.35	4.54
4	2.13	2.78	4.60	1.53	2.13	3.75
5	2.02	2.57	4.03	1.48	2.02	3.36
6	1.94	2.45	3.71	1.44	1.94	3.14
7	1.89	2.36	3.50	1.42	1.89	3.00
8	1.86	2.31	3.36	1.40	1.86	2.90
9	1.83	2.26	3.25	1.38	1.83	2.82
10	1.81	2.23	3.17	1.37	1.81	2.76
11	1.80	2.20	3.11	1.36	1.80	2.72
12	1.78	2.18	3.06	1.36	1.78	2.68
13	1.77	2.16	3.01	1.35	1.77	2.65
14	1.76	2.15	2.98	1.35	1.76	2.62
15	1.75	2.13	2.95	1.34	1.75	2.60
16	1.75	2.12	2.92	1.34	1.75	2.58
17	1.74	2.11	2.90	1.33	1.74	2.57
18	1.73	2.10	2.88	1.33	1.73	2.55
19	1.73	2.09	2.86	1.33	1.73	2.54
20	1.72	2.08	2.85	1.32	1.72	2.53
25	1.71	2.06	2.78	1.32	1.71	2.49
30	1.70	2.04	2.75	1.31	1.70	2.46
40	1.68	2.02	2.70	1.30	1.68	2.42
60	1.67	2.00	2.66	1.30	1.67	2.39
120	1.66	1.98	2.62	1.29	1.66	2.36
Infinite	1.64	1.96	2.58	1.28	1.64	2.33

Significance level

Table C Critical values for the F-test
(a) One-sided at 5% significance level

Degrees of freedom for smaller variance	Degrees of freedom for larger variance														
	1	2	3	4	5	6	7	8	9	10	12	15	20	60	Infinity
1	161.4	199.5	215.7	224.6	230.2	234.0	236.8	238.9	240.5	241.9	243.9	246.0	248.0	252.2	254.3
2	18.51	19.00	19.16	19.25	19.30	19.33	19.35	19.37	19.38	19.40	19.41	19.43	19.45	19.48	19.50
3	10.13	9.55	9.28	9.12	9.01	8.94	8.89	8.85	8.81	8.79	8.74	8.70	8.66	8.57	8.53
4	7.71	6.94	6.59	6.39	6.26	6.16	6.09	6.04	6.00	5.96	5.91	5.86	5.80	5.69	5.63
5	6.61	5.79	5.41	5.19	5.05	4.95	4.88	4.82	4.77	4.74	4.68	4.62	4.56	4.43	4.36
6	5.99	5.14	4.76	4.53	4.39	4.28	4.21	4.15	4.10	4.06	4.00	3.94	3.87	3.74	3.67
7	5.59	4.74	4.35	4.12	3.97	3.87	3.79	3.73	3.68	3.64	3.57	3.51	3.44	3.30	3.23
8	5.32	4.46	4.07	3.84	3.69	3.58	3.50	3.44	3.39	3.35	3.28	3.22	3.15	3.01	2.93
9	5.12	4.26	3.86	3.63	3.48	3.37	3.29	3.23	3.18	3.14	3.07	3.01	2.94	2.79	2.71
10	4.96	4.10	3.71	3.48	3.33	3.22	3.14	3.07	3.02	2.98	2.91	2.85	2.77	2.62	2.54
12	4.75	3.89	3.49	3.26	3.11	3.00	2.91	2.85	2.80	2.75	2.69	2.62	2.54	2.38	2.30
15	4.54	3.68	3.29	3.06	2.90	2.79	2.71	2.64	2.59	2.54	2.48	2.40	2.33	2.16	2.07
20	4.35	3.49	3.10	2.87	2.71	2.60	2.49	2.45	2.39	2.35	2.28	2.20	2.12	1.95	1.84
60	4.00	3.15	2.76	2.53	2.37	2.25	2.17	2.10	2.04	1.99	1.92	1.84	1.75	1.53	1.39
Infinity	3.84	3.00	2.60	2.37	2.21	2.10	2.01	1.94	1.88	1.83	1.75	1.67	1.57	1.32	1.00

Table C Critical values for the F-test (continued)
(b) One-sided at 1% significance level

Degrees of freedom for smaller variance	Degrees of freedom for larger variance														
	1	2	3	4	5	6	7	8	9	10	12	15	20	60	Infinity
1	4052	5000	5403	5625	5764	5859	5928	5982	6022	6056	6106	6157	6209	6313	6366
2	98.50	99.00	99.17	99.25	99.30	99.33	99.36	99.37	99.39	99.40	99.42	99.43	99.45	99.48	99.50
3	34.12	30.82	29.46	28.71	28.24	27.91	27.67	27.49	27.35	27.23	27.05	26.87	26.69	26.32	26.13
4	21.20	18.00	16.69	15.98	15.52	15.21	14.98	14.80	14.66	14.55	14.37	14.20	14.02	13.65	13.46
5	16.26	13.27	12.06	11.39	10.97	10.67	10.46	10.29	10.16	10.05	9.89	9.72	9.55	9.20	9.02
6	13.75	10.92	9.78	9.15	8.75	8.47	8.26	8.10	7.98	7.87	7.72	7.56	7.40	7.06	6.88
7	12.25	9.55	8.45	7.85	7.46	7.19	6.99	6.84	6.72	6.62	6.47	6.31	6.16	5.82	5.65
8	11.26	8.65	7.59	7.01	6.63	6.37	6.18	6.03	5.91	5.81	5.67	5.52	5.36	5.03	4.86
9	10.56	8.02	6.99	6.42	6.06	5.80	5.61	5.47	5.35	5.26	5.11	4.96	4.81	4.48	4.31
10	10.04	7.56	6.55	5.99	5.64	5.39	5.20	5.06	4.94	4.85	4.71	4.56	4.41	4.08	3.91
12	9.33	6.93	5.95	5.41	5.06	4.82	4.64	4.50	4.39	4.30	4.16	4.01	3.86	3.54	3.36
15	8.68	6.36	5.42	4.89	4.56	4.32	4.14	4.00	3.89	3.80	3.67	3.52	3.37	3.05	2.87
20	8.10	5.85	4.94	4.43	4.10	3.87	3.70	3.56	3.46	3.37	3.23	3.09	2.94	2.61	2.42
60	7.08	4.98	4.13	3.65	3.34	3.12	2.95	2.82	2.72	2.63	2.50	2.35	2.20	1.84	1.60
Infinity	6.63	4.61	3.78	3.32	3.02	2.80	2.64	2.51	2.41	2.32	2.18	2.04	1.88	1.47	1.00

Table C Critical values for the F-test (continued)
(c) Two-sided at 5% significance level

Degrees of freedom for smaller variance	Degrees of freedom for larger variance														
	1	2	3	4	5	6	7	8	9	10	12	15	20	60	Infinity
1	647.8	799.5	864.2	899.6	921.8	937.1	948.2	956.7	963.3	968.6	976.7	984.9	993.1	1010.0	1018.0
2	38.51	39.00	39.17	39.25	39.30	39.33	39.36	39.37	39.39	39.40	39.41	39.43	39.45	39.48	39.50
3	17.44	16.04	15.44	15.10	14.88	14.73	14.62	14.54	14.47	14.42	14.34	14.25	14.17	13.99	13.90
4	12.22	10.65	9.98	9.60	9.36	9.20	9.07	8.98	8.90	8.84	8.75	8.66	8.56	8.36	8.26
5	10.01	8.43	7.76	7.39	7.15	6.98	6.85	6.76	6.68	6.62	6.52	6.43	6.33	6.12	6.02
6	8.81	7.26	6.60	6.23	5.99	5.82	5.70	5.60	5.52	5.46	5.37	5.27	5.17	4.96	4.85
7	8.07	6.54	5.89	5.52	5.29	5.12	4.99	4.90	4.82	4.76	4.67	4.57	4.47	4.25	4.14
8	7.57	6.06	5.42	5.05	4.82	4.65	4.53	4.43	4.36	4.30	4.20	4.10	4.00	3.78	3.67
9	7.21	5.71	5.08	4.72	4.48	4.32	4.20	4.10	4.03	3.96	3.87	3.77	3.67	3.45	3.33
10	6.94	5.46	4.83	4.47	4.24	4.07	3.95	3.85	3.78	3.72	3.62	3.52	3.42	3.20	3.08
12	6.55	5.10	4.47	4.12	3.89	3.73	3.61	3.51	3.44	3.37	3.28	3.18	3.07	2.85	2.72
15	6.20	4.77	4.15	3.80	3.58	3.41	3.29	3.20	3.12	3.06	2.96	2.86	2.76	2.52	2.40
20	5.87	4.46	3.86	3.51	3.29	3.13	3.01	2.91	2.84	2.77	2.68	2.57	2.46	2.22	2.09
60	5.29	3.93	3.34	3.01	2.79	2.63	2.51	2.41	2.33	2.27	2.17	2.06	1.94	1.67	1.48
Infinity	5.02	3.69	3.12	2.79	2.57	2.41	2.29	2.19	2.11	2.05	1.94	1.83	1.71	1.39	1.00

Table C Critical values for the *F*-test (continued)
(d) Two-sided at 1% significance level

Degrees of freedom for smaller variance	Degrees of freedom for larger variance														
	1	*2*	*3*	*4*	*5*	*6*	*7*	*8*	*9*	*10*	*12*	*15*	*20*	*60*	*Infinity*
1	16 211	20 000	21 615	22 500	23 056	23 437	23 715	23 925	24 091	24 224	24 426	24 630	24 836	25 253	25 465
2	198.5	199.0	199.2	199.2	199.3	199.3	199.4	199.4	199.4	199.4	199.4	199.4	199.4	199.5	199.5
3	55.55	49.80	47.47	46.19	45.39	44.84	44.43	44.13	43.88	43.69	43.29	43.08	42.78	42.15	41.83
4	31.33	26.28	24.26	23.15	22.46	21.97	21.62	21.35	21.14	20.97	20.70	20.04	20.17	19.61	19.32
5	22.78	18.31	16.53	15.56	14.94	14.51	14.20	13.96	13.77	13.62	13.38	13.15	12.90	12.40	12.14
6	18.63	14.54	12.92	12.03	11.46	11.07	10.79	10.57	10.39	10.25	10.03	9.81	9.59	9.12	8.88
7	16.24	12.40	10.88	10.05	9.52	9.16	8.89	8.68	8.51	8.38	8.18	7.97	7.75	7.31	7.08
8	14.69	11.04	9.60	8.81	8.30	7.95	7.69	7.50	7.34	7.21	7.01	6.81	6.61	6.18	5.95
9	13.61	10.11	8.72	7.96	7.47	7.13	6.88	6.69	6.54	6.42	6.23	6.03	5.83	5.41	5.19
10	12.83	9.43	8.08	7.34	6.87	6.54	6.30	6.12	5.97	5.85	5.66	5.47	5.27	4.86	4.64
12	11.75	8.51	7.23	6.52	6.07	5.76	5.52	5.35	5.20	5.09	4.91	4.72	4.53	4.12	3.90
15	10.80	7.70	6.48	5.80	5.37	5.07	4.85	4.67	4.54	4.42	4.25	4.07	3.88	3.48	3.26
20	9.94	6.99	5.82	5.17	4.76	4.47	4.26	4.09	3.96	3.85	3.68	3.50	3.32	2.92	2.69
60	8.49	5.79	4.73	4.14	3.76	3.49	3.29	3.13	3.01	2.90	2.74	2.57	2.39	1.96	1.69
Infinity	7.88	5.30	4.28	3.72	3.35	3.09	2.90	2.74	2.62	2.52	2.36	2.19	2.00	1.53	1.00

Table D Critical values of the product-moment correlation coefficient

	Significance level			
	Two-sided test		One-sided test	
Sample size	5% (0.05)	1% (0.01)	5% (0.05)	1% (0.01)
4	0.950	0.990	0.900	0.980
5	0.878	0.959	0.805	0.934
6	0.811	0.917	0.729	0.882
7	0.754	0.875	0.669	0.833
8	0.707	0.834	0.621	0.789
9	0.666	0.798	0.582	0.750
10	0.632	0.765	0.549	0.715
11	0.602	0.735	0.521	0.685
12	0.576	0.708	0.497	0.658
13	0.553	0.684	0.476	0.634
14	0.532	0.661	0.457	0.612
15	0.514	0.641	0.441	0.592
16	0.497	0.623	0.426	0.574
17	0.482	0.606	0.412	0.558
18	0.468	0.590	0.400	0.543
19	0.456	0.575	0.389	0.529
20	0.444	0.561	0.378	0.516
25	0.396	0.506	0.336	0.461
30	0.363	0.464	0.307	0.423
40	0.313	0.400	0.264	0.367
60	0.254	0.331	0.215	0.300

Table E   Critical values for the chi-squared test

Degrees of freedom	Significance level	
	5% (0.05)	1% (0.01)
1	3.841	6.635
2	5.991	9.210
3	7.816	11.35
4	9.488	13.28
5	11.07	15.08
6	12.59	16.81
7	14.07	18.49
8	15.51	20.09
9	16.92	21.67
10	18.31	23.21
11	19.68	24.72
12	21.03	26.22
13	22.36	27.69
14	23.68	29.14
15	25.00	30.58
16	26.30	32.00
17	27.59	33.41
18	28.87	34.81
19	30.14	36.19
20	31.41	37.57

Table F Confidence intervals for a population standard deviation

	Level of confidence							
	90%		95%		98%		99%	
Two-sided:								
One-sided:	Lower 95% L_1	Upper 95% L_2	Lower 97.5% L_1	Upper 97.5% L_2	Lower 99% L_1	Upper 99% L_2	Lower 99.5% L_1	Upper 99.5% L_2
Degrees of freedom								
1	0.51	15.9	0.45	31.9	0.39	79.8	0.36	160 0
2	0.58	4.41	0.52	6.28	0.47	9.98	0.43	14.1
3	0.62	2.92	0.57	3.73	0.51	5.11	0.48	6.47
4	0.65	2.37	0.60	2.87	0.55	3.67	0.52	4.40
5	0.67	2.09	0.62	2.45	0.58	3.00	0.55	3.48
6	0.69	1.92	0.64	2.20	0.60	2.62	0.57	2.98
7	0.71	1.80	0.66	2.04	0.62	2.38	0.59	2.66
8	0.72	1.71	0.68	1.92	0.63	2.20	0.60	2.44
9	0.73	1.65	0.69	1.83	0.64	2.08	0.62	2.28
10	0.74	1.59	0.70	1.75	0.66	1.98	0.63	2.15
12	0.76	1.52	0.72	1.65	0.68	1.83	0.65	1.98
15	0.77	1.44	0.74	1.55	0.70	1.69	0.68	1.81
20	0.80	1.36	0.77	1.44	0.73	1.56	0.71	1.64
24	0.81	1.32	0.78	1.39	0.75	1.49	0.73	1.56
30	0.83	1.27	0.80	1.34	0.77	1.42	0.75	1.48
40	0.85	1.23	0.82	1.28	0.79	1.34	0.77	1.39
60	0.87	1.18	0.85	1.22	0.82	1.27	0.81	1.30
Infinity	1.00	1.00	1.00	1.00	1.00	1.00	1.00	1.00

Critical values for Dixon's test

	Critical values	
n	5%	1%
3	0.970	0.994
4	0.829	0.926
5	0.710	0.821
6	0.628	0.740
7	0.569	0.680
8	0.608	0.717
9	0.564	0.672
10	0.530	0.635
11	0.502	0.605
12	0.479	0.579
13	0.611	0.697
14	0.586	0.670
15	0.565	0.647
16	0.546	0.627
17	0.529	0.610
18	0.514	0.594
19	0.501	0.580
20	0.489	0.567
21	0.478	0.555
22	0.468	0.544
23	0.459	0.535
24	0.451	0.526
25	0.443	0.517
26	0.436	0.510
27	0.429	0.502
28	0.423	0.495
29	0.417	0.489
30	0.412	0.483
31	0.407	0.477
32	0.402	0.472
33	0.397	0.467
34	0.393	0.462
35	0.388	0.458
36	0.384	0.454
37	0.381	0.450
38	0.377	0.446
39	0.374	0.442
40	0.371	0.438

Table H Critical values for Cochran's test

Number of laboratories	Number of determinations in each laboratory									
	2		3		4		5		6	
	1%	5%	1%	5%	1%	5%	1%	5%	1%	5%
2	0.993		0.995	0.975	0.979	0.939	0.959	0.906	0.937	0.877
3	0.968	0.967	0.942	0.871	0.883	0.798	0.834	0.746	0.793	0.707
4	0.928	0.906	0.864	0.768	0.781	0.684	0.721	0.629	0.676	0.590
5	0.883	0.841	0.788	0.684	0.696	0.598	0.633	0.544	0.588	0.506
6	0.838	0.781	0.722	0.616	0.626	0.532	0.564	0.480	0.520	0.445
7	0.794	0.727	0.664	0.561	0.568	0.480	0.508	0.431	0.466	0.397
8	0.754	0.680	0.615	0.516	0.521	0.438	0.463	0.391	0.423	0.360
9	0.718	0.638	0.573	0.478	0.481	0.403	0.425	0.358	0.387	0.329
10	0.684	0.602	0.536	0.445	0.447	0.373	0.393	0.331	0.357	0.303
11	0.653	0.570	0.504	0.417	0.418	0.348	0.366	0.308	0.332	0.281
12	0.624	0.541	0.475	0.392	0.392	0.326	0.343	0.288	0.310	0.262
13	0.599	0.515	0.450	0.371	0.369	0.307	0.322	0.271	0.291	0.246
14	0.575	0.492	0.427	0.352	0.349	0.291	0.304	0.255	0.274	0.232
15	0.553	0.471	0.407	0.335	0.332	0.276	0.288	0.242	0.259	0.220
16	0.532	0.452	0.388	0.319	0.316	0.262	0.274	0.230	0.246	0.208
17	0.514	0.434	0.372	0.305	0.301	0.250	0.261	0.219	0.234	0.198
18	0.496	0.418	0.356	0.293	0.288	0.240	0.249	0.209	0.223	0.189
19	0.480	0.403	0.343	0.281	0.276	0.230	0.238	0.200	0.214	0.181
20	0.461	0.389	0.330	0.270	0.265	0.220	0.229	0.192	0.205	0.174
25	0.413	0.334	0.278	0.228	0.222	0.185	0.190	0.160	0.170	0.144
30	0.363	0.293	0.241	0.198	0.191	0.159	0.164	0.138	0.145	0.124
35	0.325	0.262	0.213	0.175	0.168	0.140	0.144	0.121	0.127	0.108
40	0.294	0.237	0.192	0.158	0.151	0.126	0.128	0.108	0.114	0.097

Index